Quantitative Applications in the Social Sciences

A SAGE PUBLICATIONS SERIES

1. **Analysis of Variance, 2nd Edition** *Iversen/Norpoth*
2. **Operations Research Methods** *Nagel/Neef*
3. **Causal Modeling, 2nd Edition** Asher
4. **Tests of Significance** *Henkel*
5. **Cohort Analysis, 2nd Edition** *Glenn*
6. **Canonical Analysis and Factor Comparison** *Levine*
7. **Analysis of Nominal Data, 2nd Edition** *Reynolds*
8. **Analysis of Ordinal Data** *Hildebrand/Laing/Rosenthal*
9. **Time Series Analysis, 2nd Edition** *Ostrom*
10. **Ecological Inference** *Langbein/Li*
11. **Multidimensional Scaling** *Krus*
12. **Analysis of Covariance** *Wildt/A*
13. **Introduction to Factor Analysis** *Kim/Mueller*
14. **Factor Analysis** *Kim/Mueller*
15. **Multiple Indicators** *Sullivan/Feld*
16. **Exploratory Data Analysis** *Hartw*
17. **Reliability and Validity Assessme** *Carmines/Zeller*
18. **Analyzing Panel Data** *Markus*
19. **Discriminant Analysis** *Klecka*
20. **Log-Linear Models** *Knoke/Burke*
21. **Interrupted Time Series Analysis** *McDowall/McCleary/Meidinger/Hay*
22. **Applied Regression** *Lewis-Beck*
23. **Research Designs** *Spector*
24. **Unidimensional Scaling** *McIver/Carmines*
25. **Magnitude Scaling** *Lodge*
26. **Multiattribute Evaluation** *Edwards/Newman*
27. **Dynamic Modeling** *Huckfeldt/Kohfeld/Likens*
28. **Network Analysis** *Knoke/Kuklinski*
29. **Interpreting and Using Regression** *Achen*
30. **Test Item Bias** *Osterlind*
31. **Mobility Tables** *Hout*
32. **Measures of Association** *Liebetrau*
33. **Confirmatory Factor Analysis** *Long*
34. **Covariance Structure Models** *Long*
35. **Introduction to Survey Sampling** *Kalton*
36. **Achievement Testing** *Bejar*
37. **Nonrecursive Causal Models** *Berry*
38. **Matrix Algebra** *Namboodiri*
39. **Introduction to Applied Demography** *Rives/Serow*
40. **Microcomputer Methods for Social Scientists, 2nd Edition** *Schrodt*
41. **Game Theory** *Zagare*
42. **Using Published Data** *Jacob*
43. **Bayesian Statistical Inference** *Iversen*
44. **Cluster Analysis** *Aldenderfer/Blashfield*
45. **Linear Probability, Logit, and Probit Models** *Aldrich/Nelson*
46. **Event History Analysis** *Allison*
47. **Canonical Correlation Analysis** *Thompson*
48. **Models for Innovation Diffusion** *Mahajan/Peterson*
49. **Basic Content Analysis, 2nd Edition** *Weber*
50. **Multiple Regression in Practice** *Berry/Feldman*
51. **Stochastic Parameter Regression Models** *Newbold/Bos*
52. **Using Microcomputers in Research** *Madron/Tate/Brookshire*
53. **Secondar[illegible] ey Data**

ance

Davis

rogramming

alysis

[illegible] Response *Fox/Tracy*

59. **Meta-Analysis** *Wolf*
60. **Linear Programming** *Feiring*
61. **Multiple Comparisons** *Klockars/Sax*
62. **Information Theory** *Krippendorff*
63. **Survey Questions** *Converse/Presser*
64. **Latent Class Analysis** *McCutcheon*
65. **Three-Way Scaling and Clustering** *Arabie/Carroll/DeSarbo*
66. **Q Methodology** *McKeown/Thomas*
67. **Analyzing Decision Making** *Louviere*
68. **Rasch Models for Measurement** *Andrich*
69. **Principal Components Analysis** *Dunteman*
70. **Pooled Time Series Analysis** *Sayrs*
71. **Analyzing Complex Survey Data, 2nd Edition** *Lee/Forthofer*
72. **Interaction Effects in Multiple Regression, 2nd Edition** *Jaccard/Turrisi*
73. **Understanding Significance Testing** *Mohr*
74. **Experimental Design and Analysis** *Brown/Melamed*
75. **Metric Scaling** *Weller/Romney*
76. **Longitudinal Research, 2nd Edition** *Menard*
77. **Expert Systems** *Benfer/Brent/Furbee*
78. **Data Theory and Dimensional Analysis** *Jacoby*
79. **Regression Diagnostics** *Fox*
80. **Computer-Assisted Interviewing** *Saris*
81. **Contextual Analysis** *Iversen*
82. **Summated Rating Scale Construction** *Spector*
83. **Central Tendency and Variability** *Weisberg*
84. **ANOVA: Repeated Measures** *Girden*
85. **Processing Data** *Bourque/Clark*
86. **Logit Modeling** *DeMaris*

Quantitative Applications in the Social Sciences

A SAGE PUBLICATIONS SERIES

87. **Analytic Mapping and Geographic Databases** *Garson/Biggs*
88. **Working With Archival Data** *Elder/Pavalko/Clipp*
89. **Multiple Comparison Procedures** *Toothaker*
90. **Nonparametric Statistics** *Gibbons*
91. **Nonparametric Measures of Association** *Gibbons*
92. **Understanding Regression Assumptions** *Berry*
93. **Regression With Dummy Variables** *Hardy*
94. **Loglinear Models With Latent Variables** *Hagenaars*
95. **Bootstrapping** *Mooney/Duval*
96. **Maximum Likelihood Estimation** *Eliason*
97. **Ordinal Log-Linear Models** *Ishii-Kuntz*
98. **Random Factors in ANOVA** *Jackson/Brashers*
99. **Univariate Tests for Time Series Models** *Cromwell/Labys/Terraza*
100. **Multivariate Tests for Time Series Models** *Cromwell/Hannan/Labys/Terraza*
101. **Interpreting Probability Models: Logit, Probit, and Other Generalized Linear Models** *Liao*
102. **Typologies and Taxonomies** *Bailey*
103. **Data Analysis: An Introduction** *Lewis-Beck*
104. **Multiple Attribute Decision Making** *Yoon/Hwang*
105. **Causal Analysis With Panel Data** *Finkel*
106. **Applied Logistic Regression Analysis, 2nd Edition** *Menard*
107. **Chaos and Catastrophe Theories** *Brown*
108. **Basic Math for Social Scientists: Concepts** *Hagle*
109. **Basic Math for Social Scientists: Problems and Solutions** *Hagle*
110. **Calculus** *Iversen*
111. **Regression Models: Censored, Sample Selected, or Truncated Data** *Breen*
112. **Tree Models of Similarity and Association** *James E. Corter*
113. **Computational Modeling** *Taber/Timpone*
114. **LISREL Approaches to Interaction Effects in Multiple Regression** *Jaccard/Wan*
115. **Analyzing Repeated Surveys** *Firebaugh*
116. **Monte Carlo Simulation** *Mooney*
117. **Statistical Graphics for Univariate and Bivariate Data** *Jacoby*
118. **Interaction Effects in Factorial Analysis of Variance** *Jaccard*
119. **Odds Ratios in the Analysis of Contingency Tables** *Rudas*
120. **Statistical Graphics for Visualizing Multivariate Data** *Jacoby*
121. **Applied Correspondence Analysis** *Clausen*
122. **Game Theory Topics** *Fink/Gates/Humes*
123. **Social Choice: Theory and Research** *Johnson*
124. **Neural Networks** *Abdi/Valentin/Edelman*
125. **Relating Statistics and Experimental Design: An Introduction** *Levin*
126. **Latent Class Scaling Analysis** *Dayton*
127. **Sorting Data: Collection and Analysis** *Coxon*
128. **Analyzing Documentary Accounts** *Hodson*
129. **Effect Size for ANOVA Designs** *Cortina/Nouri*
130. **Nonparametric Simple Regression: Smoothing Scatterplots** *Fox*
131. **Multiple and Generalized Nonparametric Regression** *Fox*
132. **Logistic Regression: A Primer** *Pampel*
133. **Translating Questionnaires and Other Research Instruments: Problems and Solutions** *Behling/Law*
134. **Generalized Linear Models: A United Approach** *Gill*
135. **Interaction Effects in Logistic Regression** *Jaccard*
136. **Missing Data** *Allison*
137. **Spline Regression Models** *Marsh/Cormier*
138. **Logit and Probit: Ordered and Multinomial Models** *Borooah*
139. **Correlation: Parametric and Nonparametric Measures** *Chen/Popovich*
140. **Confidence Intervals** *Smithson*
141. **Internet Data Collection** *Best/Krueger*
142. **Probability Theory** *Rudas*
143. **Multilevel Modeling** *Luke*
144. **Polytomous Item Response Theory Models** *Ostini/Nering*
145. **An Introduction to Generalized Linear Models** *Dunteman/Ho*
146. **Logistic Regression Models for Ordinal Response Variables** *O'Connell*
147. **Fuzzy Set Theory: Applications in the Social Sciences** *Smithson/Verkuilen*
148. **Multiple Time Series Models** *Brandt/Williams*
149. **Quantile Regression** *Hao/Naiman*
150. **Differential Equations: A Modeling Approach** *Brown*
151. **Graph Algebra: Mathematical Modeling With a Systems Approach** *Brown*
152. **Modern Methods for Robust Regression** *Andersen*
153. **Agent-Based Models** *Gilbert*
154. **Social Network Analysis, 2nd Edition** *Knoke/Yang*
155. **Spatial Regression Models** *Ward/Gleditsch*

Series/Number 07–155

SPATIAL REGRESSION MODELS

Michael D. Ward
University of Washington

Kristian Skrede Gleditsch
University of Essex

For information:

Sage Publications, Inc.
2455 Teller Road
Thousand Oaks, California 91320
E-mail: order@sagepub.com

Sage Publications India Pvt. Ltd.
B 1/I 1 Mohan Cooperative Industrial Area
Mathura Road, New Delhi 110 044
India

Sage Publications Ltd.
1 Oliver's Yard
55 City Road
London EC1Y 1SP
United Kingdom

Sage Publications Asia-Pacific Pte. Ltd.
33 Pekin Street #02-01
Far East Square
Singapore 048763

Printed in the United States of America

Library of Congress Cataloging-in-Publication Data

Ward, Michael Don, 1948-
Spatial regression models / Michael D. Ward, Kristian Skrede Gleditsch.
p. cm. – (Quantitative applications in the social sciences; 155)
Includes bibliographical references and index.
ISBN 978-1-4129-5415-0 (pbk. : alk. paper)
1. Spatial analysis (Statistics)
2. Regression analysis. I. Gleditsch, Kristian Skrede, 1971- II. Title.
HA30.6.W37 2008
519.5--dc22

2007037955

This book is printed on acid-free paper.

08 09 10 11 12 10 9 8 7 6 5 4 3 2 1

Acquisitions Editor:	Vicki Knight
Associate Editor:	Sean Connelly
Editorial Assistant:	Lauren Habib
Production Editor:	Melanie Birdsall
Copy Editor:	QuADS Prepress (P) Ltd.
Typesetter:	QuADS Prepress (P) Ltd.
Proofreader:	Kevin Gleason
Indexer:	Jean Casalegno
Cover Designer:	Candice Harman
Marketing Manager:	Stephanie Adams

CONTENTS

SERIES EDITOR'S INTRODUCTION

Much of quantitative social science methodology is about the analysis of individual behavior. We typically attempt to analyze individual behavior by modeling an outcome variable that measures the behavior as a function of a set of explanatory variables, most commonly in a regression type of framework. Our social science theory mainly describes how this outcome variable relates to the explanatory variables. The level of analysis, of course, does not have to be at the individual or micro level. Sometimes researchers analyze data that are at an aggregate level, such as one representing neighborhoods, communities, firms, cities, counties, states, and nations. Still, the same analytic logic applies. We seek to establish some association, causal or not, between the dependent variable and (some of) the independent variables, be the unit of analysis person, firm, community, city, state, or nation. When we do so, we implicitly assume that the geographical or spatial locations of the observations in the analysis do not matter. Even though researchers often use (dummy) variables to group together observations that are in a cluster of locations, they do so out of concern for other kinds of similarity than spatiality. For example, data analysts conventionally use a dummy variable to contrast individuals from the South in the United States against the others from the other parts of the country. They do so in an effort to capture some unique cultural traits, rather than spatial dependence in a regression analysis. This book *Spatial Regression Models* deals with precisely the spatial dependence problem in linear regression analysis.

A bit more formally, we regress the outcome variable y_i on a vector of explanatory variables $\mathbf{x}_i$ in an OLS linear regression analysis:

$$y_i = \mathbf{x}_i\beta + \varepsilon_i,$$

where β contains a vector of parameters to be estimated, and ε_i are random errors, identically and independently distributed by assumption. For the classical linear regression, the distributional assumption is normal. When there exists spatial (or other kinds of) dependence among the cases, the ε_i are no longer independent of one another, and as a result, the standard errors of β may be underestimated, thereby affecting the correctness of hypothesis testing.

Although spatial thinking in data analysis can be traced back to early attempts at cartography and surveying, modern spatial regression did not come about until recent decades when both statistical knowledge and computing power advanced. In our series, we have quite a few books either directly on the topic of linear regression or about some aspects of such analysis. However, the topic of spatial dependence has not been dealt with. Thus, Ward and Gleditsch's book is a welcome addition to the series, especially to those numbers focused on linear regression. In the book, the authors introduce the reader to two most widely used spatial regression methods, models with spatially lagged dependent variables and spatial error models, as well as some additional difficulties in spatial analysis. Although the reader's unit of analysis may not be the same as that in the authors' analyses in the book, the reader may benefit from their ample use of intuitive examples.

—Tim F. Liao
Series Editor

PREFACE

Spatial ideas can make substantial contributions to social science research. This book provides a self-contained introduction for social scientists to how the analysis of spatial dependence can be integrated into a regression framework. We intend to fill a niche, making this book accessible to a wide range of readers interested in the role of space in social science applications. Although many exhaustive surveys of spatial statistics exist, most of these are very advanced and presume that readers have a thorough knowledge of advanced statistics and probability theory (Banerjee, Carlin, & Gelfand, 2004; Cressie, 1993; Getis & Boots, 1978; Haining, 2003; Ripley, 1981, 1988; Schabenberger & Gotway, 2005). Most of these surveys are also oriented to topics and applications relevant to the natural sciences that often are unfamiliar to social scientists. We assume only that the reader is familiar with the classical regression model as widely employed in social science research and is interested in the spatial dependence that may characterize their data. In a few places we rely on matrix representations, but these are explained in considerable detail in nonmathematical language. We use the widely and freely available R computing platform (R Development Core Team, 2004) to show exactly how to implement these methods and provide some code clips to illustrate our examples. Familiarity with R at the level of Dalgaard (2002) will be useful. Other programs and approaches are also available for the analysis of spatial data, but we do not employ these in the text, although we provide brief details on some alternatives in an appendix.

This book would not have seen the light of day without support for its authors. Foremost, we thank our families for putting up with our absences during stressful times. We also thank Joan Esteban, Director of the Institute for Economic Analysis, Barcelona, Spain. Joan gave us a warm, generous welcome and enormous support as we developed the first draft of this long delayed project. Michael Ward is grateful for support he received from Adrian Raftery, the Director of the Center for Statistics and the Social Sciences (CSSS), which made this visit possible. Ward also received support from David Hodge, one-time Dean of the College of Arts and Sciences at the University of Washington, now president of Miami University, and from the chair of the Political Science Department at UW, Steve Majeski. Kristian Skrede Gleditsch received support from the National Science Foundation and a Gaspar de Portolà travel grant from the Government of Catalonia and the University of California.

Many current and former colleagues at UW, UCSD, Essex, and elsewhere have both informed and inspired our interest in dependent data. We are also grateful for insights and helpful discussions with John Ahlquist, Kristin M. Bakke, Kyle Beardsley, Nathaniel Beck, Roger Bivand, Xun Cao, Shauna Fisher, James P. LeSage, Tse-Min Lin, Michael Manger, Aseem Prakash, Andrea Ruggeri, Idean Salehyan, Michael E. Shin, Christopher Ward, Dennis L. Ward, Anton Westveld III, and Erik Wibbels. Michael Ward thanks two of his previous neighbors for their influence on his initial and continuing interest in geography: Andrew Kirby and subsequently John V. O'Loughlin occupied the office across from him in the Institute of Behavioral Science at the University of Colorado. Ward promises to return several of their books on the completion of this manuscript. Kristian Skrede Gleditsch would like to acknowledge the influence of Kidron's (1981) *State of the World Atlas* on his interest in geography and social science.

Michael Lewis-Beck encouraged one of us, in the last century, to pursue this project. He is a patient man.

—Michael D. Ward and Kristian Skrede Gleditsch
Barcelona, Seattle, San Diego, Oslo, & Colchester
October 10, 2007

ACKNOWLEDGMENTS

We are most appreciative of the constructive and helpful suggestions provided by reviewers Sudipto Banerjee, University of Minnesota, and Oliver Schabenberger, SAS.

To the accident that is geography . . . .

SPATIAL REGRESSION MODELS

Michael D. Ward
University of Washington

Kristian Skrede Gleditsch
University of Essex

1. INTRODUCTION

1.1 Interaction and Social Science

Social scientists are interested in situations in which various types of agents—individuals, political parties, groups, countries—interact with one another. In many cases, the outcomes or incentives for actions of individual actors do not depend solely on the attributes of particular individuals but on the structure of the system, their position within it, and their interactions with other individuals. Even something as prosaic as the common flu has a social component, since it is spread through social interaction. If we want to predict the likelihood that a particular individual will come down with a rhinovirus, we would look at whether something has "been going around" lately and whether the individual has been in contact with others who have become ill with this disease. Some diseases are spread via interaction, where infected individuals transmit the disease through contact with others. Clearly, different types of interaction patterns can give rise to different disease dynamics. Although demonstrably false, the spread of the HIV retrovirus to the United States is often claimed to have originated from an index case of a single Canadian airline attendant in the late 1970s (Watts, 2003).

Oddly, the role of interactions and their structures is almost completely absent from most empirical analyses in the social sciences. Consider, for example, the case of voter turnout. Differences in turnout have typically been explained using individual characteristics, such as higher education, believed to be important for political behavior. However, interaction and ties to other individuals can be as important as personal characteristics. For example, the so-called get-out-the-vote phone calls increase voting turnout by about six percentage points (±3) on average (Imai, 2005). Similarly, linkages to organizations such as churches and labor unions are also known to increase voter turnout. Baybeck and Huckfeldt (2002) show that even in dispersed networks individuals who are distant from one another are less likely to interact on a frequent basis. Such studies are the exception, not the rule; most studies of voter turnout still assume that all voters make independent decisions.

Clearly, treating observations as unrelated would be patently absurd for the flu example. Perhaps some people have weaker immune systems and are more likely to fall sick during an epidemic. However, we would not try to predict an individual's risk of the flu from his or her own attributes alone, independent of whether other individuals are infected. For example, parents are rarely "similar" to their children in terms of income, hours slept, and smoking. Yet if one is affected the other is typically also at risk. The social relations model grew out of an interest by psychologists to separate the independent and interactive effects of groups versus individuals, and it provides one attempt to model such dependencies (see, e.g., Kenny, 1981; Malloy & Kenny, 1986).

In this book, we examine how insights from *spatial* analysis can help researchers take dependence between observations into account and deal with spatially clustered phenomena. In particular, we focus on two important regression models with spatially dependent observations. The first of these concerns situations in which there is a spatially lagged dependent variable. The second focuses on spatially correlated errors. We recognize that there is a much larger set of interesting spatial modeling perspectives. This book is not intended to survey these but rather serves to introduce models with spatially lagged dependent variables and those with spatially correlated error terms. Many empirical undertakings in social science may benefit from these approaches, which have until very recently been widely ignored in the empirical social science literature. These types of models allow us to examine the impact that one observation has on other proximate observations. We believe this is important not only from first principles but also from the simple fact that many social phenomena are spatially "clustered." There are many forms of spatially organized data, ranging from geolocated individual locations for observations to regional data that are attributed to some geographical area. The latter are often called *areal* or *lattice* data, while the former are known as *point data*. In this book, we concentrate on regional data, which typically deal with units such as counties, states, provinces, and countries.[1]

Examples of spatially clustered phenomena are widespread in the social sciences. Regional voting clusters are often thought to be important in American political behavior. Political cleavages often overlap with economic and ethnic cleavages. As such, models of voter turnout may need to take into account the spatial clustering of overlapping cleavages (West, 2005). Similar examples can be found from studies in comparative politics, sociology, and economics. For example, studies of the impact of the various policy choices made by central banks have been examined for their independence from the central governments as well as the preferences of the central bankers themselves. It is widely thought that central banks are constrained by a variety of local contexts apart from how independent they are from the national authorities. Thus, even if they are independent of local authorities, are central bank policies independent of each other (Adolph, 2004; Franzese, 1999)? Murdoch, Sandler, and Sargent (1997) examine interdependent decision making in the voluntary and nonvoluntary aspects of behavior regarding emissions of sulfur dioxide and nitrous oxide in Europe during the 1980s. As pollutants are spatially dispersed without regard to national boundaries, spatial analysis techniques will help in highlighting the spillover effects of pollution as well as the interdependence of compliance issues. Inequality and poverty are thought to be intertwined in cross-national studies. The most skewed wealth and income distributions are often in the poorest countries. Recent work has shown that corruption is often the consequence, as well as plausible cause, of poverty. However, it turns out that income inequality may increase the level of corruption, even more so than poverty. It may be that the distributions of wealth and corruption share a spatial clustering that complicates this effect. Spatial analysis can help untangle this conundrum. Recent work along these lines includes You and Khagram (2005). Finally, organizational forms may also spread in much the same way—policy emulation. Holmes (2006) addresses the contagion of unionism with spatial models.

In short, there are myriad studies across the gamut of the social sciences that employ data that are actually organized on a spatial template, whether the units are counties, cities, states, countries, or firms. It often turns out that the characteristics of these units are highly clustered in particular spatial regions. In many of these applications, it is plausible to assume that there may be dependencies across the observations. In practice, this clustering is generally ignored or treated as a nuisance. Ignoring these dependencies imposes a substantial price on our ability to generate meaningful inferences about the processes we study. Spatial analysis provides one way of reducing that price and taking advantage of the information we have about how social processes are interconnected. We next turn to a simple example of how this works in an important area of social science—namely, the study of the diffusion of democratic institutions.

1.2 Democracy Around the World

To motivate our discussion, we use a simple example with data where observations are unlikely to be independent of one another. Social scientists have long been interested in possible explanations for why some countries are democracies and others not. An early and influential contribution by Lipset (1959) suggested that there were social requisites for democratic rule. One of these requisites was high levels of average income; Lipset noted that "the average wealth . . . is much higher for the more democratic countries" (p. 75). This argument—which has served as a cornerstone of comparative analysis for more than four decades—suggested that societies with higher average income were more likely to have democratic institutions. Table 1.1 provides an abridged list of data on gross domestic product (GDP) per capita income and level of democracy for several countries in the world in 2002. Our measure of democracy is the so-called POLITY index, which classifies countries on a series of institutional criteria. The index ranges from –10 for the least democratic societies to 10 for the most democratic societies. Gleditsch and Ward (1997) provide further details on the construction of this index. We have sorted Table 1.1 on GDP per capita and democracy so that it is easier to see simple patterns among the variables. As can be seen, some wealthy societies, such as Denmark, are indeed democratic, while low-income countries, such as Sierra Leone and North Korea, are autocracies. Interestingly, Lipset suggested that in 1959 Australia, Belgium, Canada, Denmark, Ireland, Luxembourg, the Netherlands, New Zealand, Norway, Sweden, Switzerland, the United Kingdom, and the United States comprised the list of "stable democracies" in Europe, North America, and South America. In 1959, the unstable democracies and dictatorships included Austria, Finland, France, West Germany, Italy, and Spain. Most of these are now democracies and generally considered stable. Despite looking at some cases that clearly are consistent with Lipset's claim, is there a strong general relationship between wealth and democracy? India is democratic in spite of low average national income, and although India has recently experienced high rates of growth, it remains far below the levels observed for Organization for Economic Cooperation and Development countries. At the same time, it is also hard to ignore the existence of many autocracies situated in the Middle East that have relatively high incomes, which seems to contradict the claim made by Lipset. To evaluate the relationship more generally, we turn to a systematic comparative analysis.

TABLE 1.1
GDP Data for 2002

Country	*Democracy*	*GDP*	*Country*	*Democracy*	*GDP*
Guinea	–1	51	Iran	3	1,776
Ethiopia	1	114	Macedonia	6	1,801
Burundi	0	120	Namibia	6	1,870
Zaire	0	135	Romania	8	1,941
Sierra Leone	–10	172	Algeria	–3	2,036
Eritrea	–7	175	Bosnia & Herzegovina	0	2,108
Malawi	5	178	Thailand	9	2,215
Iraq	–9	181	Suriname	9	2,224
Guinea-Bissau	5	187	Guatemala	8	2,257
Liberia	0	194	Russia	7	2,279
Rwanda	–4	216	Ecuador	6	2,305
Mozambique	6	217	Peru	9	2,306
Tajikistan	–1	221	Colombia	7	2,342
Niger	4	247	Jordan	–2	2,375
Nepal	6	276	Fiji	5	2,397
Burkina Faso	0	315	Tunisia	–4	2,436
Chad	–2	317	El Salvador	7	2,486
Uganda	–4	320	South Africa	9	2,607
Tanzania	2	330	Dominican Republic	8	2,745
C. African Rep.	5	333	Cuba	–7	2,891
⋮	⋮	⋮	⋮	⋮	⋮
Turkmenistan	–9	1,241	Canada	10	25,139
Morocco	–6	1,300	Finland	10	26,235
Congo	–5	1,303	Austria	10	26,304
Djibouti	2	1,313	Netherlands	10	27,059
Belarus	–7	1,359	Sweden	10	27,497
North Korea	–9	1,361	United Kingdom	10	27,650
Swaziland	–9	1,412	Japan	10	31,731
Albania	5	1,416	United Arab Emirates	–8	34,436
Syria	–7	1,417	Qatar	–10	36,611
Kazakhstan	–6	1,437	Denmark	10	37,063
Serbia	7	1,573	Switzerland	10	39,769
Egypt	–6	1,602	United States	10	40,180
Myanmar (Burma)	–7	1,729	Norway	10	43,895
Bulgaria	9	1,744	Luxembourg	10	54,255

NOTE: GDP figures are per capita. An abridged list is shown; full data are available on the Web site for the volume.

Following the work of Lipset (1959) and many others since, it is common in empirical, comparative work on democracy to consider democracy as a linear function of the natural log of GDP per capita. We estimate the level of democracy in a country, measured by the POLITY score, given its GDP per capita using ordinary least squares (OLS) regression:

$$\text{POLITY score} = \beta_0 + \beta_1 \ln \text{GDP per capita} + \epsilon.$$

The estimates for this linear regression of democracy on GDP per capita are shown in Table 1.2. The positive sign of the coefficient for ln GDP per capita illustrates the positive relationship between democracy and income, but the estimated impact is relatively small when we take into account the metric of the variables.

More specifically, this linear model predicts that a country with Uzbekistan's GDP per capita ($464 in 2002) would have a democracy score of approximately 1. In contrast, for a country that has a level of GDP per capita income approximately twice that of Uzbekistan ($1,020), the model predicts an associated democracy score of about 2. For most analysts, scores of 1 and 2 are considered to be similar on the POLITY democracy index. Thus, there does not seem to be a large impact of even fairly dramatic differences in income on the predicted level of democracy, despite the statistical significance of the estimated coefficient for the log of GDP per capita.

Figure 1.1 shows that the estimated OLS equation predicts democracy levels of poor countries that are far higher than their actual levels. Nonetheless, the implied, estimated effect of wealth on democracy is not only small—more than doubling the GDP per capita has a small impact on democracy—for poor countries, such as Uzbekistan, but probably overestimated as well. Almost any standard analysis of these residuals will reflect the first impression given in this figure: They do not look "well behaved," in the sense that we have two peaks of observations around high and low values where the model underpredicts or overpredicts the actual

TABLE 1.2
OLS Estimates of Democracy as a Linear Function of Logged GDP Per Capita

	$\hat{\beta}$	$SE(\hat{\beta})$	*t Value*
Intercept	–9.69	2.43	–3.99
ln GDP per capita	1.69	0.31	5.36
$N = 158$			
Log likelihood ($df = 3$) = –513.62			
$F = 28.77$ ($df_1 = 1$, $df_2 = 156$)			

NOTE: Estimates obtained using 2002 data from the POLITY project and the World Bank.

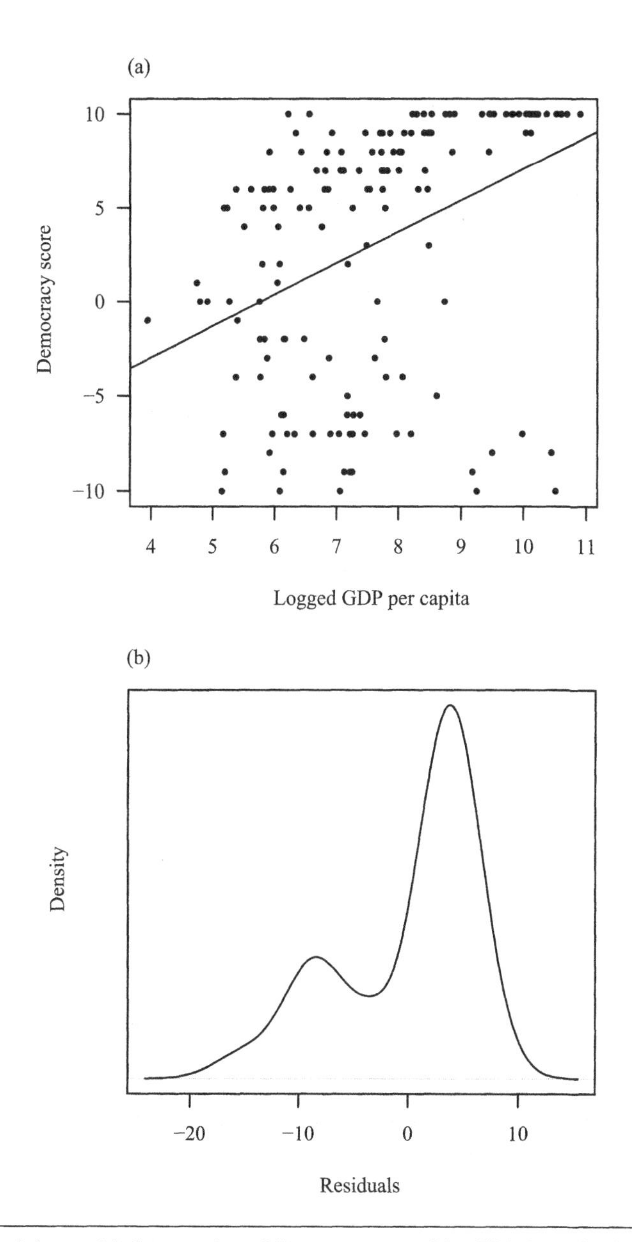

Figure 1.1 (a) Scatterplot of Democracy and ln GDP Per Capita With Regression Line; (b) Densityplot of Residuals From OLS Regression. These Suggest "Clustering" of Autocracies at the Lower End and Democracies at the Higher End.

level of democracy. Figure 1.1 also shows substantial and patterned variation around the estimated regression line or general tendency. But are these residuals organized in a way that is dependent on the interdependencies of the observations? Panel (b) in Figure 1.1 shows convincingly that the residuals are not distributed normally, nor are they even unimodal. Instead, there is a cluster of negative values around –10 and a cluster of positive values around 5. Thus, it is clear in this example that the residuals from the OLS regression reported in Table 1.2 are problematic and raise issues about whether the estimated coefficients in that regression can be trusted. These residuals suggest that the underlying systematic model does not capture the relationship between democracy and economic output very well, probably in part as a result of dependencies among the data—specifically clustering of similar values. It may be that countries exert influence on each other in ways that would produce such results.

1.3 Introducing Spatial Dependence

One possibility for explaining these results is that, in addition to characteristics of individual countries, the prospects for democracy in one country are not independent of whether neighboring countries have democratic institutions or not. During the Cold War, Soviet intervention enforced socialist rules in many states in Eastern Europe. Moreover, democratic transitions in many Latin American states appear to have been influenced by processes in other countries (see, e.g., Gleditsch & Ward, 2007). Looking at the data in Table 1.1 organized alphabetically, it would be hard to identify easily whether there are any pockets or regions of similar regimes beyond what we would expect from GDP per capita. Even with the information sorted on salient features for comparison, careful analytical study may be required to identify various kinds of patterns.

Exploratory examination of plausible spatial (and spatial-like) clustering may be important in a variety of situations, revealing aspects of social interaction that are missing from unconnected displays. Potentially unobserved clusters can influence our understanding of what is actually occurring in the part of the model we think we *do* understand. Before we turn to an examination of how to take spatial correlation into account, we explain a bit more about why it is important to do so.

Even if an analyst simply wants to compare means and construct classical statistical tests, such as difference of means tests, problems arise if the data are spatially correlated. Consider a one-sample t test on variable y defined as

$$t = \frac{1/n\sum_{i=1}^{n} y_i}{\sigma/\sqrt{n}}.$$

If there is a correlation among observations that are near one another temporally or spatially (first-order serial correlation), then the actual standard error will be larger for positive values of serial correlation (and smaller for negative values). Researchers tend to be sensitive to the problem of serially correlated observations over time but often neglect the fact that the same problem will apply for serial correlation across observations at the same point in time. Using the unadjusted estimate of the variance will result in having a *t* value that is larger than warranted. This increases the chance of making a Type I error, even for situations in which there is only a small amount of spatial autocorrelation and abundant observations.

In short, because of serial, spatial correlation among the observations—or whatever reason—classical tests are biased in terms of accepting the hypothesized substantive account, even when it is untrue. Assuming that the data are spatially dependent such that the dependence is inversely proportional to the distance between observations, ρ represents the resultant first-order spatial correlation. This correlation measures how similar neighbors are on some measured attribute. As a result of this correlation, the true standard error for the mean is given approximately by

$$\sigma_{\bar{y}} \approx \sqrt{\frac{1+\rho}{1-\rho}}\frac{\sigma}{\sqrt{n}}.$$

A simple way to understand the impact of spatial correlation is to imagine a variable *y* observed on *n* observations: $y_1, y_2, \ldots, y_{n-1}, y_n$. In many situations, we think of these observations as being independent of one another and each identically distributed, typically from a normal distribution of unknown mean μ and variance σ^2. The typical estimator of μ is

$$\bar{y} = \sum_{i=1}^{n} y_i / n.$$

Since the observations are thought to arise from a normal distribution, inference is based on y and σ. The 95% confidence interval is given as $\bar{y} \pm 1.96\sigma/\sqrt{n}$. If there is spatial correlation among the y_i, that is, greater similarity the closer observations y_i and y_j are to each other spatially, then, as Cressie (1993, p. 14) shows, the covariance for positive values of ρ will be

$$\text{cov}(y_i, y_j) = \sigma^2 \times \rho^{|i-j|},$$

and the variance is

$$\operatorname{var}(\bar{y}) = n^{-2}\left\{\sum_{i=1}^{n}\sum_{j=1}^{n}\operatorname{cov}(y_i, y_j)\right\},$$

which expands to

$$\operatorname{var}(\bar{y}) = \left\{\frac{\sigma^2}{n}\right\}\left[1+2\left\{\frac{\rho}{1-\rho}\right\}\left\{1-\frac{1}{n}\right\}-2\left\{\frac{\rho}{1-\rho}\right\}^2\frac{1-\rho^{n-1}}{n}\right].$$

The factor

$$\left[1+2\left\{\frac{\rho}{1-\rho}\right\}\left\{1-\frac{1}{n}\right\}-2\left\{\frac{\rho}{1-\rho}\right\}^2\frac{1-\rho^{n-1}}{n}\right]$$

essentially is the discount on the number of observations that is imposed by spatial correlation, which does not disappear in large samples. If $n = 10$ and $\rho = .26$ (as in Cressie's example), then the discount is about 40%: 10 spatially correlated observations have the same precision as about 6 independent observations. This, in turn, implies that ignoring the spatial correlation leads to a confidence interval that is far too small when there is positive spatial correlation among observations. *In general, ignoring spatial dependence will tend to underestimate the real variance in the data.* Thus, for a sample of 158 observations on GDP, the 95% confidence band under an assumption of normality would be $(1.96\times\sigma)/\sqrt{n}$, but if there were a spatial correlation of 0.65—the actual value of $\hat{\rho}$ for GDP from the above example—the correct confidence interval would be approximately 4.22 instead of 1.96, over twice as large. In the case of the level of democracy, $\hat{\rho}$ is 0.47, which leads to a 95% confidence band that is $(3.26\times\sigma)/\sqrt{n}$, which is almost 70% wider.[2]

If there are different forms of spatial correlation, then different specific adjustments may be required, but the general point is that if there is positive spatial correlation the sample mean will have less precision. As a result, the null hypothesis will frequently be rejected when it is true. Thus, it is unwise to rely on statistical tests that perform well in independent and identically distributed (iid) samples if the underlying data are spatially (inter)dependent. Schabenberger and Gotway (2005) illustrate this relative excess variability of the least squares estimator for different levels of autocorrelation in different sample sizes. For $\rho > 0$, this excess variability rises with n such that with $\rho = .9$ the excess variability is approximately 14.0 when the sample size approaches 50. The important point is that spatially correlated data will wreak considerable havoc with statistical tests designed for iid data, leading researchers to reject the null hypothesis because the standard tests underestimate the variability.

1.4 Maps as Visual Displays of Data

Humans are adroit at recognizing patterns, even where no patterns exist. Often, this is where statistics comes in. However, in a heuristic, exploratory model, it is useful to know everything there is to know about your data. Dense tables of numerical information are important ways of conveying a great deal of information, albeit slowly. Graphical displays provide an auxiliary method that may allow patterns to be discovered visually, quickly. However, it is important to use graphical techniques in the context of a plausible explanation of the phenomena of interest. Recent work has illustrated the importance of the careful display of evidence and quantitative material, as well as providing a gold standard (Cleveland, 1993; Tufte, 1990, 1992, 1997; Wainer, 2004). One guiding principle is that the method of display needs to bear a strong relationship to the explanatory story being developed.

The classic study of the spread of cholera in mid–19th century London by John Snow, popularized in Tufte (1997) and more recently and completely elaborated by Johnson (2006), provides a good example of such a geographical story. Snow demonstrated that the spread of the main outbreak of cholera in London during the summer of 1854 was a result of Soho inhabitants (and others) drinking water from a pump on Broad Street, which had become infected from the burial site of many of the victims of the cholera epidemic. Thus, proximity to the Broad Street well was a potent risk factor for cholera and this fact played an important role in rejecting the theory that cholera was airborne. Snow's maps of London have become classics illustrating how spatial correlation can embody causal thinking. Figure 1.2 provides the classic map of the Soho district that illustrates that many cholera deaths were clustered around the Broad Street pump.

Shaded maps are also an important way for displaying processes that have a geographical story. In our example, we are suggesting that there is feedback among proximate countries that influences their political institutions and economic wealth. Figure 1.3 provides a world "map" shaded by level of democracy measured in 2002 for 158 countries. The story of this map is that democratic institutions are clustered in neighboring countries and that autocratic countries are also clustered in different regions of the world. Countries on this map are shaded in increasing levels of gray

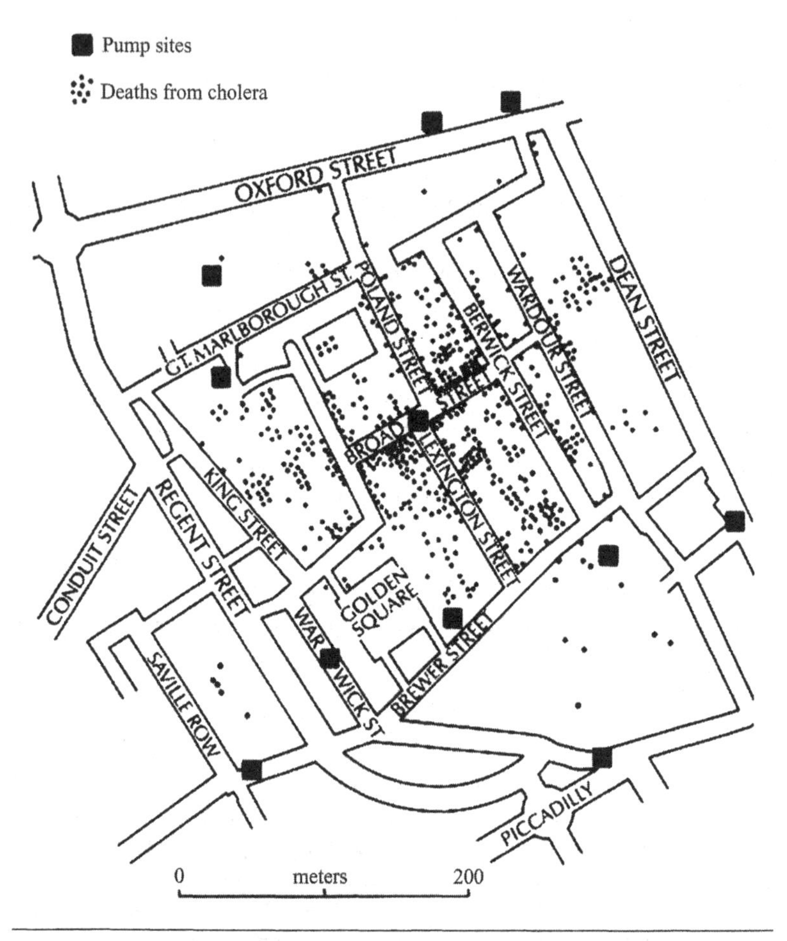

Figure 1.2 John Snow's Map of Cholera Deaths in the Soho District of London During the Summer of 1854.

for higher levels of democracy. Only countries with the highest democracy score are shaded in black; France, for example, has the next lightest shade owing to its coding as a 9 on the democracy scale to reflect the relative independence of the President from the National Assembly. Figure 1.3 shows that there are strong geographical clusters of democracies and autocracies. By and large, most of the democratic countries are located in Western Europe and the Americas, along with Australia and Oceania, while many autocratic countries are to be found in Africa, the

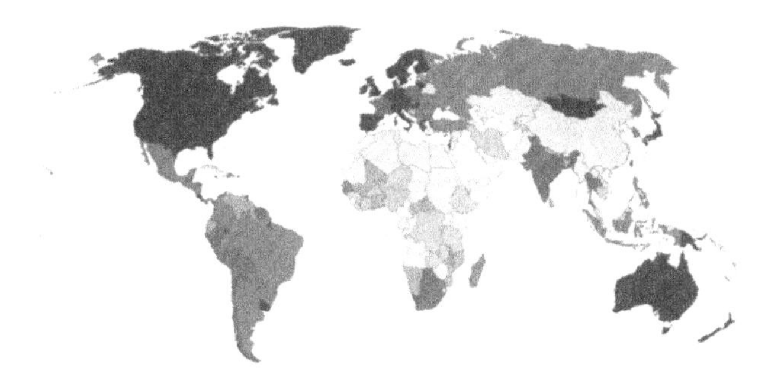

Figure 1.3 Countries With High Levels of Democracy, as Measured by the POLITY IV Indicators, Are Shown in Darker Shades of Gray and Tend to Be Grouped Together Around the World. Similarly, Countries Rated as More Authoritarian Tend to Be Grouped Geographically.

Middle East, and Asia. There are of course exceptions, with countries such as Belarus persisting as an autocracy surrounded by (mostly) more democratic neighbors. In contrast, India is a democracy surrounded by mostly nondemocratic neighbors.

In Latin America, most states are democracies in 2002, despite large differences in their GDP per capita. By comparison, in the Middle East, most states are autocratic, despite having GDP per capita levels that are consistently higher than the world average. Indeed, mapping these attributes suggests that both democracy and GDP per capita display spatial clustering. In many cases, visualization and mapping reveal structure in the data that is not readily available from looking at the data in tabular format.

Figure 1.4 illustrates the clustering of GDP per capita (logged) in 2002. Wealthy countries are colored in darker shades of gray, while poorer ones are shown in lighter shades. Figure 1.4 also presents a picture of strong clustering. North America and Western Europe are clusters of wealthy countries, while especially Africa is composed of countries that are poor and also have poor neighbors. There are, of course, exceptions in wealth as well. Japan and Australia, for example, are on average much richer than their neighbors.

Cartographic data displays are great tools for exploratory spatial data analysis, but good displays will always embody an empirical or theoretical story.

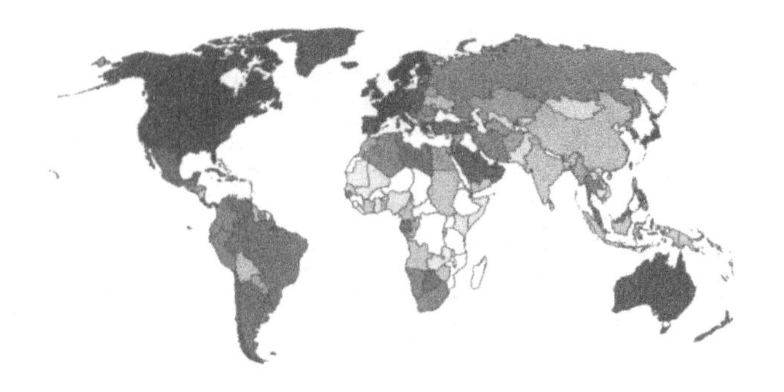

Figure 1.4 Countries With High Levels of GDP Per Capita Are Shown in Darker Grays.

1.5 Measuring Spatial Association and Correlation

Unfortunately, just as patterns may be ignored in a data matrix, humans are adept at seeing structure when there really is none. As such, it is useful to have more formalized ways of evaluating whether observations are spatially clustered or related across some forms of ties between observations. We turn to formal exploratory tools in the next section.

Exploring such associations, however, requires that we have some idea about which observations are likely to be related to one another. For a set of n units, each observation i can be potentially related to all the $(n-i)$ possible units, but in practice, however, we can usually assume that some interactions or ties are more important than others. The network or structure between units that we are interested in must generally be specified *prior* to analysis of dependence between other observations. The techniques that we explore here usually start from a graph or list L of relations between connected observations. For many purposes, it is practical to use a matrix to represent the connectivities between observations. For example, we can define a binary matrix $\mathbf{C}$ that specifies connectivities between individual observations. We have an entry $c_{ij} = 1$ if two observations i and j are considered connected, $c_{ij} = 0$ if not.

The basic ideas of measuring spatial associations and correlations can be thought of as cross-product statistics, following Hubert, Golledge, and Constanzo (1981), which cross-multiply a measure of spatial proximity with a measure of the similarity of values on some particular attribute.[3] Let S_{ij} be some measure of the spatial proximity of two observations i and j and let U_{ij} be the similarity on some underlying variable of concern. Cross-product statistics will have the *general* form

$$\sum_{i=1}^{n}\sum_{j=1}^{n} S_{ij}U_{ij} \quad \forall\, i \neq j.$$

If U_{ij} defines similarity as a mean-normalized cross-product on the underlying variable, say $\left[(y_i - \overline{y})(y_j - \overline{y})\right]$, then with appropriate scaling, summing this product over all observations yields a measure of spatial correlation known as the Moran $\mathcal{I}$ statistic. If U_{ij} is defined as a squared difference, such as $(y_i - y_j)^2$, the resulting statistic is known as Geary's $\mathcal{C}$. In this book, we primarily focus on Moran's $\mathcal{I}$.[4]

For example, spatial association in the case of measures of democracy would join a measure of how close countries were to one another in terms of some spatial measurement, such as whether they had borders within 200 km of one another, with a measure of the similarity of democracy scores for each pair of countries examined. These statistics are useful as heuristics for identifying spatial patterns. Perhaps they are most useful as a diagnostic heuristic for examining the residuals from modeling exercises in which it is believed there is no (remaining) spatial patterning not accounted for by the model used.

The first task in formally assessing such correlations is to specify the interdependencies among data. This requires developing a list of which observations are connected to one another. This is an important step, but one that we will only illustrate here. Linkages might be established by physical distance, say the distance between capital cities. However, other transmission mechanisms such as the density of transportation networks via roadways, trains, waterways, and air carriers may be a better indicator of connection in particular circumstances. Similarly, instead of capital city distances, scholars have used the length of the border between neighboring countries, for example, as a measure of interaction opportunities among adjacent countries. In Gleditsch and Ward (2001), we develop a database of the minimum distances among all countries in the world. We use these data herein, specifying that countries are neighbors if they have a minimum distance of 200 or fewer kilometers between them.[5]

A subset of these data is portrayed in Table 1.3 in two ways, first as a list and then as a matrix. Many computer programs organize large matrices as lists, since it allows a more efficient storage of information, allowing only the nonzero elements to be included in memory. Indeed, for small subsets, it is easier, perhaps, to derive spatial characteristics and record them as lists

TABLE 1.3
Connectivity Matrix for a Subset of European Countries

List Format

Country	*Connections*
Denmark	Germany, Norway, Sweden
Finland	Norway, Sweden
France	Germany, Italy, UK
Germany	Denmark, France, Italy, Sweden
Italy	France, Germany
Norway	Denmark, Finland, Sweden
Sweden	Denmark, Finland, Germany, Norway
UK	France

Connectivity Matrix Format

	Denmark	*Finland*	*France*	*Germany*	*Italy*	*Norway*	*Sweden*	*UK*
Denmark	0	0	0	1	0	1	1	0
Finland	0	0	0	0	0	1	1	0
France	0	0	0	1	1	0	0	1
Germany	1	0	1	0	1	0	1	0
Italy	0	0	1	1	0	0	0	0
Norway	1	1	0	0	0	0	1	0
Sweden	1	1	0	1	0	1	0	0
UK	0	0	1	0	0	0	0	0

NOTE: A connection is present if countries have borders within 200 km of one another.

of connections. However, each list can be converted easily into a square matrix that portrays the observations along the rows and columns and the linkages in the interior of the matrix. A matrix representation is also helpful for defining certain variables or measures reflecting spatial structures and variation. The first part of Table 1.3 presents a set of connectivity data as a list; the second part illustrates the corresponding binary connectivity matrix **C** of these connections.

These data can also be presented as a simple network graph, as in Figure 1.5. Such graphs are illuminating, but they quickly become convoluted, crowded, and difficult to read when the number of nodes is high. Panel (b) shows the crowding in the network map of all 158 countries. However, such visual network representations may be a useful way to examine some data sets, especially those that are smaller.

Once we have a potential network of connections between observations specified by a list L or a connectivity matrix **C**, we can explore

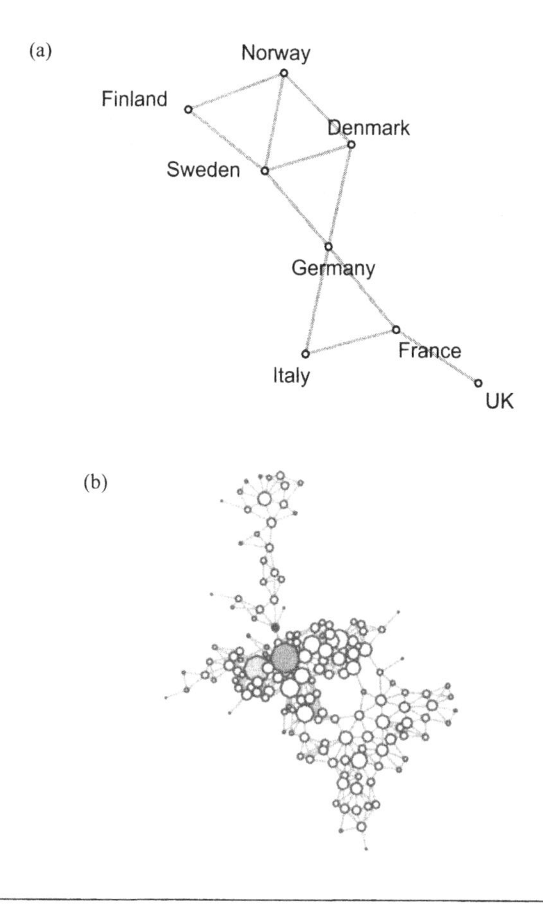

Figure 1.5 A Simple Network Representation of the Data in Table 1.3 and Among 158 Countries: (a) Linkages Among Eight European Countries; (b) Linkages Among 158 Countries.

NOTE: The United States is black and China and Russia are shades of gray in Panel (b). Size of nodes is proportional to the number of countries within 200 km.

whether the values on a variable of concern, which we denote here as y, are similar across connected or neighboring observations. One way to do this would be to look at whether two connected observations i and j tend to be similar to one another—for example, by determining whether high or low values for i tend to go together with high or low values for j.

But i is usually connected to many observations and we do not have spatial clustering unless it is similar to many of its neighbors. To combine information about the connected observations, we usually assume that all neighbors carry equal weight and that the weight of each is proportional to 1 over the total number of connectivities. The main goal of getting a "spatial lag" is to derive an average value that exists in the neighboring region. What is the average value of democracy in the neighbors of the United States? What is the average value of GDP per capita of Ghana's neighbors? Are these average values of neighboring observations correlated with each country's own score on democracy or GDP per capita? We present a heuristic statistic for gauging this, a statistic that measures the spatial correlation. In much the same way that a researcher might generate the correlation matrix among independent variables, this spatial correlation might also provide heuristic information about the observed data.

Let y_i^s denote the mean or average of y across all connected observations, or the "lag" of y over space. Matrix representation makes it easier to see the construction of the spatial lag y_i^s from y and the connectivity matrix **C**. We can create a row-normalized connectivity weight matrix **W** where each row sums up to 1 by dividing each row vector $c_{i.}$ of the binary connectivity matrix **C** by the total number of links $\sum c_{i.}$. An example is given in Table 1.4.

In this context, the scalar $y_i^s = c_{i.}y$ calculates (by summing) the average or mean across all neighboring observations of one unit i. This is often referred to as the *spatial lag*. The relationship $y^s = \mathbf{W}y$ reminds us of how

TABLE 1.4
Row-Standardized Connectivity Matrix
for a Subset of Eight European Countries

	Denmark	*Finland*	*France*	*Germany*	*Italy*	*Norway*	*Sweden*	*UK*
Denmark	0	0	0	⅓	0	⅓	⅓	0
Finland	0	0	0	0	0	½	½	0
France	0	0	0	⅓	⅓	0	0	⅓
Germany	¼	0	¼	0	¼	0	¼	0
Italy	0	0	½	½	0	0	0	0
Norway	⅓	⅓	0	0	0	0	⅓	0
Sweden	¼	¼	0	¼	0	¼	0	0
UK	0	0	1	0	0	0	0	0

NOTE: A connection is present if countries have borders within 200 km of one another.

TABLE 1.5
Ten Countries With the Largest and Smallest Spatial Lags

Country	*Democracy*	*Spatial Lag*
Largest negative spatial lags		
Bahrain	–8	–10
Tajikistan	–1	–7.1
Oman	–9	–6.7
Kyrgyzstan	–3	–6.6
United Arab Emirates	–8	–6.5
Uzbekistan	–9	–6
Qatar	–10	–5.8
Yemen	–2	–5.5
Kuwait	–7	–5.3
Israel	10	–5
Largest positive spatial lags		
Luxembourg	10	9.8
Switzerland	10	9.8
United Kingdom	10	9.8
Belgium	10	9.8
Netherlands	10	9.8
Canada	10	10
Fiji	5	10
France	9	10
Ireland	10	10
Portugal	10	10

NOTE: Countries are listed with their corresponding democracy and spatially lagged democracy scores.

each y_i^s is related to values of y for other states and the connectivity weights $\mathbf{w}_i$. Table 1.5 presents the 10 largest positive and negative spatial lags for the democracy variable. Bahrain has a democracy score of –8, for example, but is surrounded by neighboring countries that all have the maximum negative democracy score, –10. Ireland and Portugal, on the other hand, have the highest possible democracy score as do all of their neighbors.

1.6 Measuring Proximity

For many social scientists, developing a measure of the proximity of units being studied is perhaps the most important step in spatial analysis.

What is distance, in a social context? While many physical scientists will be able to use a strict measure of geographical or Euclidean distance to gauge how close trees are to one another, for example, this issue is considerably more complicated for many social science analyses. How close are, for example, the United States and Mexico? If we use a strict contiguity measure, they are perfect neighbors since they share a land border. But Canada also shares a land border with the United States. Does this imply that it is equally close to the United States? The straight-line distance from Washington, D.C., to Mexico City is approximately 3,000 km, while the distance from Washington, D.C., to Ottawa is about 700 km. We might use the length of borders between countries or the distances between the average of the 10 largest population centers in each country. Figure 1.6 illustrates the difference between these two specifications. In some countries, the centroid (open dots) is quite distant from the actual capital city (black dots), but in small countries this cannot be the case. China, Canada, Russia, Australia, and the United States are examples that illustrate the distance between these two locations. In contrast, in North and South Korea there is little distance between the centroids and the capital cities.

Another important issue in applied work is how to deal with missing spatial data. Imputation may be one approach, though other alternatives exist (Griffith, 2003). A real problem is that social science data are frequently missing, but rarely randomly missing. In nonspatial applications, this may be handled in the standard fashion—by imputation or, more frequently, by deletion of observations with missing information. However, in the spatial framework, such missing data may create "holes" in the spatial representation and undermine establishing a salient and complete representation of the spatial proximities. Another problem that can occur in some kinds of spatial setups is that some observations will not be linked to other observations. For example, New Zealand is not within 200 km of any other independent country. Two strategies are widely employed to circumvent these situations. Island isolates are often deleted from the analysis, since at a substantive level they are not "connected" and thereby will not affect other observations via the spatial process being studied. More prosaically, deleting them will purge the resulting spatial weights matrix of certain singularities (rows and columns composed entirely of 0s). A second strategy is simply to choose the "nearest" or most plausible neighbors for the islands, linking Australia and New Zealand as neighbors, for example, even if all other linkages are set for 200 km. More generally, one can use nearest k neighbor distances for all units.

Figure 1.6 can be generated in R via the following commands:

```
# Set working directory
dd <- c("C:...")
setwd(dd)

# Plotting map with centroids and capitals

# Load required libraries
library(RColorBrewer);library(maptools)
library(spdep);library(sp);library(rgdal)

# Read a Robinson projection map from an ESRI shapefile
rob.shp <- read.shape("wg2002worldmap.shp")

# Indicate the id codes for each polygon/country
rob.map <- Map2poly(rob.shp,region.id =
     unique(as.character(rob.shp$att.data$FIPS_CNTRY)))

# Indicate the map projection
tr <- readShapePoly("wg2002worldmap", IDvar="FIPS_CNTRY",
     proj4string=CRS("+proj=robin +lon 0=0"))

# Extract the relevant variables and exclude missing data
ct <- na.omit(rob.shp$att.data[,c(1, 18:20)])

# Assign relevant variable/column names
colnames(ct) <- c("ID","x", "y", "City_POP")
ct$x <- as.numeric(as.character(ct$x))
ct$y <- as.numeric(as.character(ct$y))

# Add coordinates
coordinates(ct) <- c("x", "y")
proj4string(ct) <- CRS("+proj=longlat +datum=WGS84")

# Transform the coordinates to the robinson projection
ct_rb <- spTransform(ct, CRS=CRS("+proj=robin +lon0=0"))

# Replot the map itself without a bounding box
plot(rob.map, border="Grey", forcefill=T, xaxt="n", yaxt="n",
     bty="n", lwd=.000000000125, las=1, ylab="",
     main="Centroids and Capitals", xlab="")

# Add the centroids
points(coordinates(tr), pch=19,cex=.5, col="grey")

# Add the capitals
points(coordinates(ct_rb), pch=19, cex=.5, col="black")
```

```
# Add segments between centroids and capitals
tr_or <- coordinates(tr)
rownames(tr_or) <-
     as.character((attributes(tr)$data)$FIP S_CNTRY)
ct_rb_or <- coordinates(ct_rb)
rownames(ct_rb_or) <- as.character(ct_rb$ID)

# Delete Kiribati (91), as longitude extends across
     international date line
coor_dif <- cbind(tr_or[-91,], ct_rb_or[rownames(tr_or),][-91,])
x1 <- coor_dif[,1]
x2 <- coor_dif[,3]
y1 <- coor_dif[,2]
y2 <- coor_dif[,4]
segments(x1, y1, x2, y2,col="slategray4")
```

Above we have suggested two basic metrics for measuring distance, but this just scratches the surface. This metric of distance could be thought of in terms of average travel times, the number of mobile phone conversations between each pair of points, the amount of tourism from each point to every other location, or any variety of different measures of distance and interactions. Countries that have a large amount of commerce with each other, for example, can be thought of as economically "close" (Lofdahl, 2002). Griffith (1996) offers some ideas about how such measures can and should be developed.

It would seem natural to estimate the similarity between states' own level of democracy and the levels of their neighbors by the correlation between y

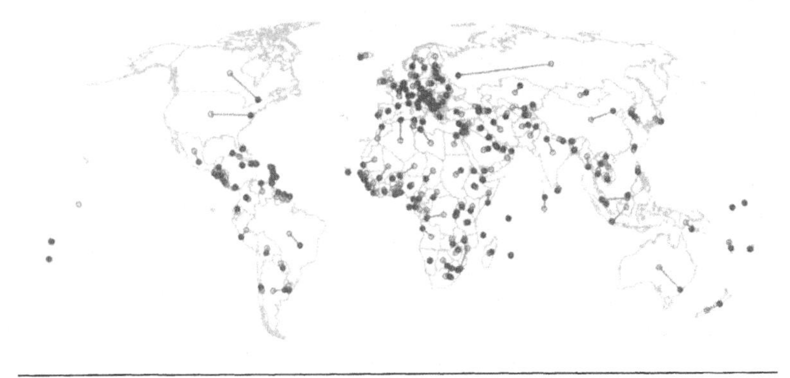

Figure 1.6 Map With Paths Between Geographical Centroid (Open Dots) and Capital City (Black Dots).

and y^s. The linear association between a value and a weighted average of its neighbors is known as Moran's $\mathcal{I}$ statistic (Moran, 1950a, 1950b), a global correlation of the values of an observation with those of its neighbors. The generalized Moran's $\mathcal{I}$ is given by a weighted, scaled cross-product:

$$\mathcal{I} = \frac{n\sum_i \sum_{j \neq i} w_{ij}(y_i - \bar{y})(y_j - \bar{y})}{\left(\sum_i \sum_{j \neq i} w_{ij}\right)\sum_i (y_i - \bar{y})^2},$$

where w denotes the elements of the row-standardized weights matrix $\mathbf{W}$ and y is the variable of concern.

$\mathcal{I}$ can be considered normal (asymptotically) with a mean that is $-1/(n-1)$. The variance of Moran's $\mathcal{I}$ is then given by

$$\operatorname{var}(\mathcal{I}) =$$

$$\frac{n^2(n-1)\frac{1}{2}\sum_{i \neq j}(w_{ij} + w_{ji})^2 - n(n-1)\sum_k\left(\sum_j w_{kj} + \sum_i w_{ik}\right)^2 - 2\left(\sum_{i \neq j} w_{ij}\right)^2}{(n+1)(n-1)^2\left(\sum_{i \neq j} w_{ij}\right)^2}.$$

If the variable of concern is standardized as z_i, Moran's $\mathcal{I}$ is simply

$$\mathcal{I} = \frac{1}{2}\sum_{ij} c_{ij} z_i z_j \quad \forall\, i \neq j.$$

Moran's $\mathcal{I}$ statistic is often used as a test of spatial correlation by constructing a z score with the mean and variance components.

Moran's $\mathcal{I}$ does not really have a fixed metric, and its expected value is $-1/(n-1)$ rather than 0. However, Moran's $\mathcal{I}$ statistic can be given a graphical interpretation that helps convey how spatial association among individual cases will give rise to different values of the statistic. Consider a scatterplot of $\tilde{y}$ against its average among neighbors' $\tilde{y}^s$ (we use a standardized $\tilde{y} = (y - \bar{y})/sd(y)$ so that the value has a mean of 0 and a standard deviation of 1). In this plot, the distribution of observations in the four quadrants around the mean of $\tilde{y}$ and $\tilde{y}^s$ captures a picture of the spatial association of the variable y. If there is no spatial clustering or association in y, the individual values of y^s should not vary systematically with y. However, if there is a positive spatial association, individual observations that have values above or below the mean on y should also be low and high, respectively, on y^s, or among proximate countries. The bulk of the cases should fall in the south-west and north-east quadrants where units are similar to their neighbors, and we should have few observations in the north-west or south-east quadrants. If we fit a regression line to this scatterplot, its slope is the Moran $\mathcal{I}$ correlation given the original variable y and the connectivity list L or matrix $\mathbf{C}$.

Figure 1.7 provides a stylized plot illustrating the Moran $\mathcal{I}$ statistic and the interpretation of a scatterplot of a variable and the first-order spatial lag. The slope of the regression line is the average spatial correlation in the data; it is the Moran $\mathcal{I}$ statistic.

Moran's $\mathcal{I}$ compares the relationship between the deviations from the mean across all neighbors of i, adjusted for the variation in y and the number of neighbors for each observation. Higher values of Moran's $\mathcal{I}$ indicate stronger positive (geographical) clustering; that is, values for neighboring units are similar to one another. This statistic measures the average correlation of an observation with its neighbors. Figure 1.7 illustrates the basic concept. The spatial lag (the average value of one's neighbors) is shown on the vertical axis, while the horizontal axis portrays the value of each

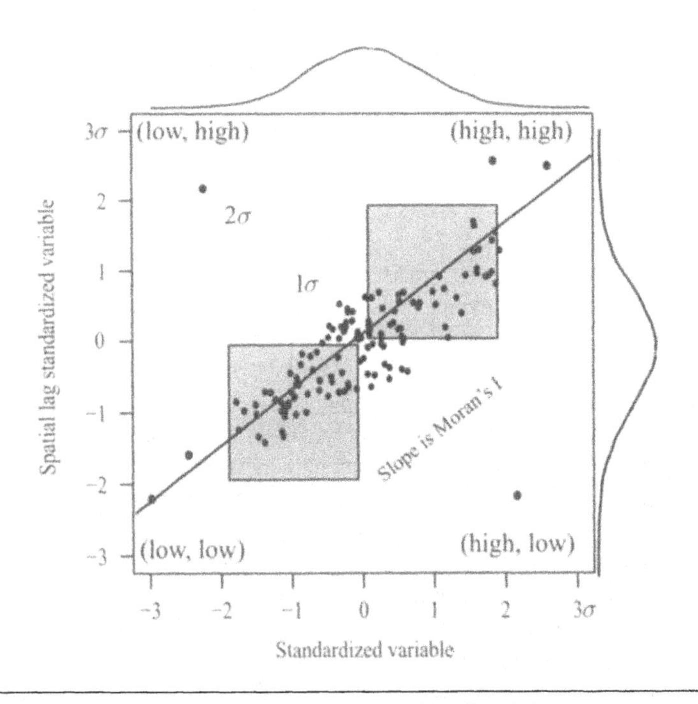

Figure 1.7 Scatterplot of a Variable and Its Spatial Lag.

NOTE: The variable is standardized to have a mean of 0 and a variance of 1. Clusters of observations in homogeneous neighborhoods are shown in shaded areas. The OLS regression line is also plotted.

observation, standardized to have means of 0 and variances of 1. A box is drawn to indicate ± 2, and most of the observations fall inside this boundary (note that $2\sigma = 2$ here since the variable is standardized). Those observations that fall in the shaded boxes are cases that are in homogeneous neighborhoods. Those in the upper, shaded area between (0, 0) and (2, 2) are observations that are above the mean on the measured variable *and* have neighbors that, on average, are also above the mean. Similarly, the shaded box between (0, 0) and (–2, 2) are observations that are below the mean and also have neighbors similarly characterized. This diagonal has many of the observations in this scatterplot and highlights the clustering of similar values. But there are a few cases that represent observations with low values on the observed variable, but with neighbors that are, on average, much higher than the mean on this variable. The single point in the upper left part of the scatterplot is such an observation, one that can be thought of as an enclave of low values in a neighborhood of high ones. An OLS regression line through all these standardized observations will produce a summary measure of the relationship between the value of an observation and that held by its neighbors. If, for example, the variable of concern were crime rates and these were measured for each precinct in metropolitan Houston, the observations in the upper right part of the figure would be precincts with high crime rates, surrounded by precincts that also had high crime rates. Similarly, those in the lower left would represent precincts with low crime rates, surrounded by precincts that also had low rates of crime. The slope of the regression line through these standardized points is the Moran $\mathcal{I}$ statistic.

Statistical testing of the Moran $\mathcal{I}$ coefficient requires additional assumptions, since we need to have the first and second moments (mean and variance) for simple probability testing in the classical framework. First and foremost is the often overlooked step of the hypothesis-testing framework: specifying the null hypothesis. For a spatial model, this is not necessarily obvious as there are many patterns that are substantially different from any spatially organized variable. For example, is the spatial pattern normally distributed? Or, is it randomly distributed? If random, exactly how is it random in space? Generally, two approaches are found in the literature, each somewhat ad hoc. The first of these assumes that the data are normally distributed. Cliff and Ord (1971) worked out the variance of $\mathcal{I}$ under this condition. There is substantial work suggesting that assuming a normal distribution for Moran's $\mathcal{I}$ is an often incorrect assumption (Boots & Tiefelsdorf, 2000; Tiefelsdorf, 1972), but most software and applied articles still assume normality.[6] The second approach is to use Monte Carlo simulation to randomly permute the rows and columns of the connections matrix sufficiently to provide a randomized null model. Both of the main

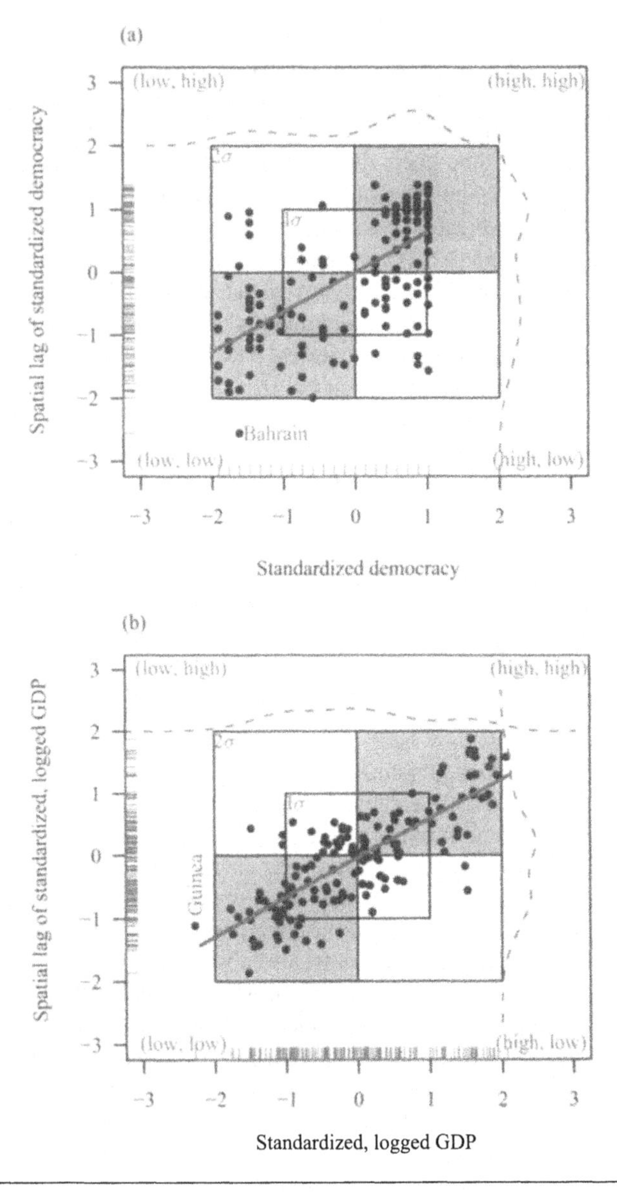

Figure 1.8 Plot of Standardized Variable Against Its Spatial Lag: (a) Democracy, Moran's $\mathcal{I} = 0.64$; (b) GDP Per Capita, Moran's $\mathcal{I} = 0.65$.

options are available in most software and will generally—although not always—yield similar results.

The standardized scatterplot for democracy is given in Figure 1.8a. This graphic has a box drawn at $\pm 2\sigma$ to give a sense of which observations are vastly unusual. Observations that occur in the top right quadrant represent those cases that have relatively high values, "surrounded" with other cases that also have high values. Cases in the bottom left quadrant are cases that have relatively low values and are surrounded by other cases with similarly low values, Bahrain being the extremum. The "off-diagonal" cases represent cases that, in this example, are surrounded by countries with vastly different levels of democracy. As illustrated, there are remarkably few such cases for autocracies (the most exceptional being Belarus) and even fewer for democracies. The figure also portrays a regression line, the slope of which provides the Moran $\mathcal{I}$ for democracy (0.64) that is much larger than the expected value of the statistic in this example (–1/158). The plot for GDP per capita is given in Figure 1.8b. Guinea is the country at the lower left, outside the 2σ square; Luxembourg is just outside this area at the top right.

We can use Moran's $\mathcal{I}$ on the OLS residuals from the estimation reported in Table 1.2 to see if the residual variation appears to display spatial clustering.[7] This is only a heuristic check, much like using it on the raw data themselves. This can be easily done in R, using the defined regression object `ols1.fit` and a list of neighbors, defined using a 200-km distance band from the outer boundaries of countries to determine each country's "neighbors," contained in the list `nblist`.

```
source("chapter1data.R")
ols1.fit <- glm(democracy ~ log(gdp.2002/population), data=sldv)
library(spdep)        # Load spdep library for moran.test()
moran.test(resid(ols1.fit),nb2listw(nblist))
lm.morantest(ols1.fit,nb2listw(nblist))
```

The computed Moran $\mathcal{I}$ statistic for these OLS residuals is 0.40, with a variance of 0.0028. This has an associated standard score of 7.77 that is much larger than $-1/158$ and has an associated p value that is ≈ 0. This tells us that the OLS results, which assume independent observations, are strongly affected by the spatial clustering in the dependent and independent variables. As a result, they are likely to be misleading for both the statistical and substantive inferences that we may wish to draw about the relationship between democracy and its social requisite of wealth, as captured in GDP per capita.

1.7 Estimating Spatial Models

What might constitute a simple set of steps for spatial analysis?

First: Map the data, especially the dependent variable. This can be done in a variety of contexts, ranging from spreadsheet plugins, map mashups, to GIS packages, but we find it best to undertake this in the context of a platform that will permit statistical analysis of these data. We illustrate the use of R libraries, especially `maptools` and `spdep`, for constructing simple maps of the distribution of variables.

Second: Also, determine if there is some discernable spatial correlation in the dependent variable. For most applications—that is, not point processes—that we consider in this book, this means calculating the Moran $\mathcal{I}$ statistic to gauge the magnitude of spatial correlation. Analysts may in some cases wish to proceed to examine and plot/map each observation's contribution to spatial correlation, through local indicator of spatial association (LISA). We do not pursue this in any detail in this book. See Gleditsch and Ward (2000), Anselin (1995), and Ord and Getis (1995) for further discussion and examples.

Third: Precisely incorporate these spatially lagged variables into the basic statistical framework and examine the resultant residuals for remaining spatial association.

Fourth: In addition to employing the normal model heuristics to gauge the fit of the model and the degree of uncertainty in the estimated parameters, the equilibrium impact should be computed and examined. This means teasing out the equilibrium, feedback implications of the estimated spatial model for the dependent variable.

We now turn to an illustration of these steps in the context of our running example.

1.7.1 Mapping the Data and Constructing the Spatial Weights Matrices

We have illustrated mapping of data with the democracy scores for 158 countries in 2002. In this subsection, we illustrate the use of mapping with the residuals from the OLS model. The data themselves were mapped earlier in Figures 1.3 and 1.4. We have also shown that the residuals from the

Figure 1.9 Geographical Display of Residuals From the OLS Regression.

regression of democracy on income display spatial association. We calculated Moran's $\mathcal{I}$ using a 200-km distance band from the outer boundaries of countries to determine each country's "neighbors." As previously reported, the Moran $\mathcal{I}$ in this case had a value of 0.40 and a variance of 0.0028. This is significant in a classical sense and allows us to be confident that the spatial patterns that were perceived in Figures 1.3 and 1.4 are actually influencing the regression results in a substantial fashion, that is, introducing bias into estimates and standard errors. These residuals from the OLS are shown in Figure 1.9.

1.7.2 Looking for Spatial Patterns

We also illustrate the construction of a so-called Shin spatial scatterplot, adapted from the work of Shin (2001). This plots the standardized value of each input variable—in this case the residuals—against its spatial lag or average value for its connected observations (Figure 1.10). The shaded boxes indicate concordant observations where a value above the mean of the residual is accompanied by a positive value for its neighbors. The axes contain a "rug plot," indicating the distribution of the variables. An estimated kernel density estimate of the distribution of the variable itself and the spatial is displayed in the outer margins. We provide the code to generate this plot:

```
pdffilename <- c("file name and path")
pdf(file=pdffilename, width=5.0, height=5.0, family="Times")
dem <- (resid(ols1.fit)) # residuals
ds <- (dem-mean(dem))/sqrt(var(dem)) # standardized democracy score
```

```
# create spatial lag and standardize it
ds.slag <- as.vector(wmat%*%ds)
ds.slag <- (ds.slag-mean(ds.slag))/sqrt(var(ds.slag))
plot(ds,ds.slag,xlim=c(-3,3),ylim=c(-3,3),pch=20,las=1,
         xlab="standardized democracy",
         ylab="spatial lag of standardized democracy")
reg1 <- lm(ds.slag~ds)

# establish a grid
xgrid <- seq(-3,1.5,length.out=158)
x0 <- list(ds=xgrid)
pred.out<-predict(reg1,x0,interval="confidence")

# put 1 and 2 sigma boxes on plot
lines ( c(-2,-2,+2,+2,-2),c(-2,+2,+2,-2,-2))
lines ( c(-1,-1,+1,+1,-1),c(-1,+1,+1,-1,-1))
lines ( c(-2,+2),c( 0, 0))
lines ( c( 0, 0),c(-2,+2))

# some text for context
text(-2.5,3,"(low,high)");text(2.5,3,"(high,high)")
text(-2.5,-3,"(low,low)");text(2.5,-3,"(high,low)")
polygon(x=c(-1,0,0,-1), y=c(-1,-1,0,0), col = "slategray3")
polygon(x=c(0,1,1,0), y=c(0,0,1,1), col = "slategray3")

# plot c.i. region
polygon(x=c(xgrid,rev(xgrid)), y=c(pred.out[,3],
         rev(pred.out[,2])), col="slategray3", border=T)

# put data on plot
points(ds,ds.slag,pch=20)

# densities
sldensity <- density(ds.slag)
lines(sldensity$y+2,sldensity$x,lty=2,col="slategray4")
ddensity <- density(ds)
lines(ddensity$x,ddensity$y+2,lty=2,col="slategray4",xlim=c(-2,2))
points(ds,ds.slag,pch=20)
lines(xgrid,pred.out[,1],type="l",lty=2,col="gray80",lwd=2)

# rugs on two sides
rug(jitter(ds,factor=2),col="slategray3")
rug(ds.slag,side=2,col="slategray3")

# label some points
text(-2.,-2.3,"Oil Exporters",col="slategray4")
dev.off()
```

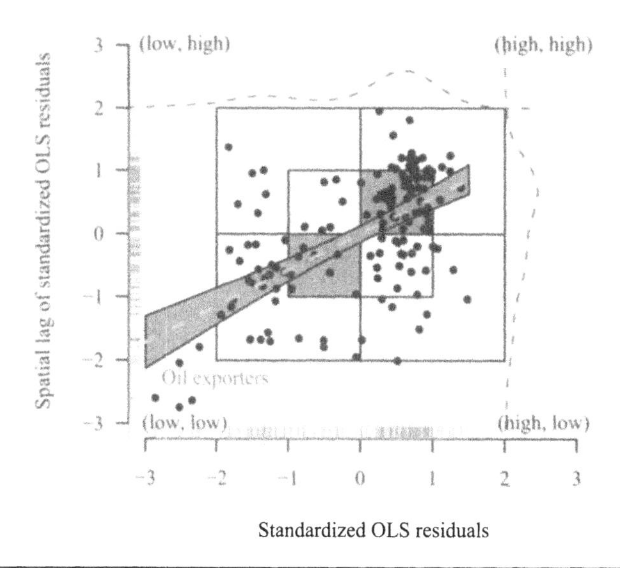

Figure 1.10 Shin Plot of OLS Residuals.

Next we turn to examining the spatial associations of the measurements on democracy. The spatial lag of the democracy variable is simply the average level of democracy in surrounding countries. Countries having neighbors with high democracy scores will have a high value here, and countries with more autocratic neighbors will have large negative values. We map these in Figure 1.11. The map illustrates that most of the countries

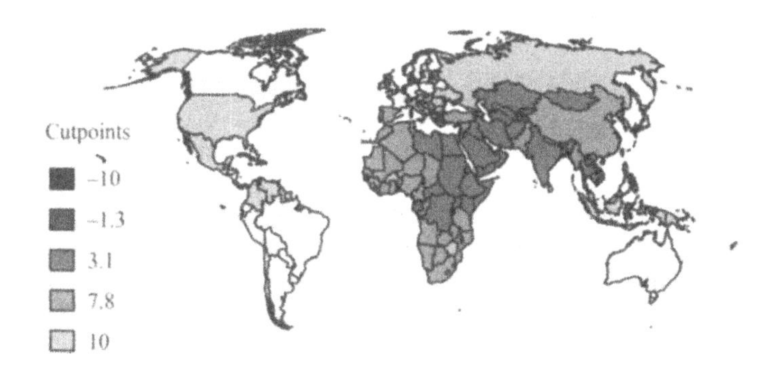

Figure 1.11 The Spatial Lag of Democracy; Darker Shades Indicate Higher Values on the Spatial Lag Variable.

in Africa and Asia are characterized by nondemocratic neighborhoods, while Europe and much of the Americas have countries with democratic neighbors (Gleditsch, 2002a).

The code to implement this is as follows:

```
# sldv2 is the data.frame
# mdd2 is the minimum distance data.frame
nblist <- vector(mode="list",length=dim(sldv2)[1])
attr(nblist,"region.id") <- sldv2$tla
attr(nblist,"class") <- "nb"
nbnms <- data.frame(sldv2$tla,c(1:dim(sldv2)[1]))
names(nbnms) <- c ("acr","nm")

min200 <- mdd2[mdd2$mindist<=200,] # Create an index of the isolates
nodata <- setdiff(sldv2$tla,unique(c(min200$ida,min200$idb)))

# Find neighbors for each row in the sldv for(i in 1:dim(sldv2)[1]){
   temp <- min200[min200$ida==sldv2$tla[i] |
           min200$idb==sldv2$tla[i],]
   cty <- unique(c(temp$ida,temp$idb))
   cty <- setdiff(cty,sldv2$tla[i])
   nblist[[i]] <- nbnms[match(cty,nbnms$acr),"nm"]
}

# wmat is the row standardized weights matrix
wmat <- matrix(0,ncol=dim(sldv2)[1],nrow=dim(sldv2)[1])
rownames(wmat) <- colnames(wmat) <- sldv2$tla
for (i in 1:dim(min200)[1]){
    wmat[min200$ida[i],min200$idb[i]] <- 1
}

wmat <- wmat/rowSums(wmat)

# calculate the spatial lag of democracy
democracy.spatial.lag <- as.vector(wmat%*%sldv2$democracy)
```

In addition to mapping the first-order spatial lag of democracy, it is also useful to map the contributions of each observation to the global Moran $\mathcal{I}$ statistic. This quantity is known as the LISA statistic. Herein we standardize these and provide a mapping that is displayed in Figure 1.12. The local Moran is developed in Ord and Getis (1995), Anselin (1995), and Getis and Ord (1996).

This map illustrates which countries have the most unusual situations in terms of their neighbors' level of democracy. Southern and western Africa

Figure 1.12 z Scores From the Local Moran's $\mathcal{I}$ Statistics Are Shown in Gray Scale Representing Similarity to Neighbors.

fall into this category, as does India. However, we note that China has a large positive score on the Moran $\mathcal{I}$, while several of its neighbors are at the opposite end with large negative Moran $\mathcal{I}$ scores.

1.8 Summary

Having first carefully examined our data and visual displays of these data, we explored the results of an OLS regression positing that the level of democracy is a linear function of wealth, measured as logged GDP per capita. We inspected the residuals from this regression and found convincing evidence that the residuals appear to display spatial clustering, violating the regression assumption that the error terms of individual observations can be considered independent of one another. As such, OLS assuming independent observations will not be a compelling method for analyzing the relationship between income and democracy. More fundamentally, a model assuming independent observation where only income matters for democracy ignores important features of obvious geographical clustering. We have also shown how maps and simple statistics can be used as informative heuristics to assess the extent and nature of spatial clustering.

Even if one is not interested in regression analysis, there is room for examining spatial patterns in social science data. We show that whether one is going to simply do a test of means or use a regression approach to examining data that are spatially organized, failure to take the spatial correlation into account will lead to incorrect inferences that are generally biased away from rejecting the stated hypotheses.

Cartographic displays of correlational data provide an exploratory heuristic for determining the presence of spatial patterns, patterns that can complicate statistical inference. We turn next to estimation of regression models with spatially lagged dependent variables, an approach that can help take spatial dependencies explicitly into account within a regression framework.

Notes

1. There are very few applications of point data in the social sciences, as yet. A recent exception is Cho and Gimpel (2007).

2. Even for the mean, Grenander (1954) illustrates that the minimum unbiased estimator should not ignore the value of correlated observations:

$$\hat{\mu} = \frac{\left[y_1 + (1-\rho)\sum_{i=2}^{n-1} y_i + y_n\right]}{\left[n-(n-2)\rho\right]}.$$

3. In the context of spatial point processes, these are sometimes known as join-count statistics, since they count the number of neighboring points with similar joint attributes.

4. Both Geary and Moran made important contributions in a number of areas. Geary is most well-known for the Stone-Geary utility function, as well as the methods for calculating the purchasing power parity for international comparisons of real income.

5. These data are available at http://privatewww.essex.ac.uk/~ksg/mindist.html.

6. Moran's $\mathcal{I}$ is developed for looking at single variables. When there is an explicit multivariate model, such as the case with OLS residuals, it is recommended that one use a slightly different version of the Moran statistic, to guard against overestimating the extent of spatial correlation. Practically, there is often little difference, and the widespread practice is to use the standard Moran statistic in both these instances. Tiefelsdorf (1972) developed a saddlepoint approximation to the underlying distribution, and this method does a better job than Moran's $\mathcal{I}$ at matching the tails of the distribution in many situations. We are indebted to Roger Bivand for his thoughtful advice on this matter. The R `spdep` function `lm.morantest.sad()` implements the Tiefelsdorf saddlepoint approach. A more straightforward approach is to use Lagrange multiplier tests to test for specific forms of spatial autocorrelation, an approach we expound in the subsequent chapter.

7. As noted by Haining (2003, pp. 276–283), the standard Moran $\mathcal{I}$ may overstate the spatial correlation of residuals if there is strong spatial patterning of the independent variables.

2. SPATIALLY LAGGED DEPENDENT VARIABLES

In this chapter, we describe a statistical model that incorporates spatial dependence explicitly by adding a "spatially lagged" dependent variable y on the right-hand side of the regression equation. This model goes by many different names. Anselin (1988) calls this the *spatial autoregressive* model, but this terminology is potentially confusing since the term *autoregressive* is used to denote quite different spatial models in the geostatistical literature. For simplicity, we will here call it the spatially lagged y model, since its main feature is the presence of a spatially lagged dependent variable among the covariates.

The spatially lagged y model is appropriate when we believe that the values of y in one unit i are directly influenced by the values of y found in i's "neighbors." This influence is above and beyond other covariates specific to i. If we believe that y is not influenced directly by the value of y as such among neighbors but rather that there is some spatially clustered feature that influences the value of y for i and its neighbors but is omitted from the specification, we may consider an alternative model with spatially correlated errors, which we discuss subsequently. For the spatially lagged y model to be appropriate, the dependent variable y must be considered as a continuous variable. In this book, we do not examine the generally more complicated case of binary dependent variables. These are more complicated since they often do not have a closed-form solution and must be estimated with iterative techniques outside the range of this book (see Ward & Gleditsch, 2002).

2.1 Regression With Spatially Lagged Dependent Variables

To motivate and illustrate the spatially lagged y model, we return to our example of the distribution of democracy around the world. We have seen that the distribution of democracy displays spatial clustering in the sense that countries are more likely to have higher values on the POLITY democracy score if they are surrounded by countries that also have high levels of democracy. Although some of the clustering in democracy obviously could be due to spatial clustering in GDP per capita, which in turn is positively related to democracy, we have shown that the spatial clustering in the democracy data does not completely disappear when we condition on a country's level of GDP per capita. The assumption that the errors ϵ_i of a model treating democracy as a function of GDP per capita are independent can easily be examined by testing for possible spatial dependence in the residuals from the regression; that is, $\hat{\epsilon}_i = (\hat{y}_i - y)$, using the Moran $\mathcal{I}$ correlation coefficient and our specified pattern of connectivities in the matrix **C**, where states in this case are defined as connected if they are within a 200-km distance threshold of one another. In this instance, we found strong evidence of residual spatial correlation. The Moran $\mathcal{I}$ statistic for these residuals is 0.40, which has an associated z score of approximately 8.[1] This is far from what we would expect if the null hypothesis of spatial independence were true. Stated differently, this implies that there is considerable positive association between the democracy level of a country and that of its geographical neighbors, above and beyond what we would expect from their levels of GDP per capita. This result is fairly typical, and it will often be the case that including spatially clustered covariates alone will not completely remove spatial clustering in the outcome of interest.

Given that the distribution of democracy still displays spatial clustering after conditioning on a country's GDP per capita, we should look for possible ways to incorporate this spatial dependence in our previous regression model. As in the case of serial clustering over time, we can think of spatial autocorrelation either as nuisance or substance. Spatial dependence leads to problems with the regression estimate $\hat{\beta}$ for the effect of GDP per capita and its standard errors, since the errors cannot be considered to be independent among connected units. These problems in estimating the effect of GDP per capita on democracy can, in principle, be addressed through alternative estimators that take into account the spatial correlation of the errors, that is, the residual variation not captured by GDP per capita alone. This approach is often known as the spatial error model, an approach we discuss subsequently.

However, our broader interest here is in what influences democracy, not just estimating the association between a country's GDP per capita on its prospects for democracy. If a country's level of democracy appears to be associated with its neighbors' level of democracy, this tells us something important about the distribution of democracy itself and provides an opportunity for learning something about possible influences from spatial dependence on prospects and constraints on democracy. As such, a more plausible and interesting approach is to consider the spatial association as a substantive feature of democracy rather than as a statistical nuisance.

The spatial association observed here suggests that we have dependence among observations such that the expected value of democracy for a country i differs notably depending on the level of democracy in neighboring states j. Instead of letting expected democracy for a country i depend just on GDP per capita, we devise a model where democracy is a function of both its own GDP per capita and the level of democracy among neighbors, defined by $w_{i\cdot}y_i$ where the entries of the connectivity vector $w_{i\cdot}$ (i.e., row i from matrix **W**) acquire nonzero values for all states j that are defined as connected to i. Recall again that the **W** connectivity matrix is row standardized so that each row in **W** sums to 1.

This reasoning suggests a spatially lagged dependent variable model of the form

$$y_i = \beta_0 + \beta_1 x_i + \rho w_{i\cdot} y_i + \epsilon_i, \qquad [2.1]$$

where a positive value for the parameter associated with the spatial lag (ρ) indicates that countries are expected to have higher democracy values if, on average, their neighbors have high democracy values.

One can think of the spatially lagged y model as analogous to an autoregressive time series model where temporal serial correlation is addressed by including a lagged dependent variable y_{t-1} on the right-hand side when we estimate the effects of other right-hand side covariates (say x_t) on y_t. The $\hat{\beta}_1$ coefficient in the spatially lagged y model differs from the coefficient calculated via the OLS regression model in that we are now assessing the effect of GDP per capita on the democracy level of a country, while controlling for spatial dependence in y, or the extent to which variation in a country i's level of democracy can be accounted for by the value of y in other connected countries j. Hence, we will need to take into account the spatial ramifications when assessing the effect of changes in x.

Tables 2.1 and 2.2 provide the estimates from an OLS regression on the level of democracy on the natural log of per capita GDP in 158 countries in 2002 with and without a spatial lag of y. We observe a large positive coefficient for the log of GDP per capita of 1.68 in the OLS without the spatially lagged y. In contrast, in the spatially lagged y model, the estimated coefficient for the log of GDP per capita is 0.76, less than half of its original size,

TABLE 2.1
OLS Without Spatial Lag

OLS	$\hat{\beta}$	*SE*($\hat{\beta}$)	*t Value*
Intercept	–9.69	2.43	–3.99
ln GDP per capita	1.68	0.31	5.36
$N = 158$			
Log likelihood ($df = 3$) = –513.62			
$F = 28.77$ ($df_1 = 1$, $df_2 = 156$)			

TABLE 2.2
OLS With Spatial Lag

OLS	$\hat{\beta}$	$SE(\hat{\beta})$	*t Value*
Intercept	–4.98	2.07	–2.40
ln GDP per capita	0.76	0.28	2.72
ρ	0.76	0.088	8.65
$N = 158$			
Log likelihood ($df = 4$) = –482.48			
$F = 58.64$ ($df_1 = 2$, $df_2 = 155$)			

although it continues to be far away from 0 by the conventional standards for significance tests.

The estimate for the spatially lagged *y* term is large and positive (0.76) and highly statistically significant by standard criteria. This provides support for the conjecture that a country's level of democracy covaries with the level of democracy among its geographical neighbors. In substantive terms, the model implies that a country's expected level of democracy would be –7.6 points lower if its neighbors had an average democracy score at the minimum possible score (i.e., –10) compared with a neighbor average of 0, which is close to the historical average POLITY score since 1945. Conversely, a country with the maximum neighbor average democracy score of 10 would be expected to be 7.6 points "more democratic" relative to a country with a neighbor average of 0. These estimates reflect the clustering of democracy illustrated previously. Although most democracies tend to have higher GDP per capita, we do observe clusters of democracy in 2002 in areas where GDP per capita is not particularly high, as in Latin America, and clustering of autocracies in areas with high average GDP, as in the Gulf states.

Comparing the measures of the overall fit for the model assuming independent observations in Table 2.1 and the model with the spatially lagged *y* in Table 2.2 indicates that the model with the spatially lagged *y* term fits the data notably better. It has a higher *F* statistic and a higher log likelihood than the model assuming independent observations. This, in turn, reinforces our belief that the spatial lag of *y* adds something important to specifying the distribution of democracy, beyond what we would expect from a country's GDP per capita. Model heuristics alone do not provide the compelling reason for using the spatial approach, however. The spatial approach is better not because of the heuristics it produces alone, but because it specifies a plausible form of the feedback or dependency among observations.

A standard ordinary least squares regression has the following form:

$$y_i = \mathbf{x}_i \beta + \varepsilon_i.$$

If ε_i is decomposed into a spatially lagged term for the dependent variable—which is correlated with the dependent variable—and an independent error term, $\varepsilon_i = \rho \mathbf{w}_{i\cdot} y_i + \epsilon_i$, this leads to the formulation for spatially lagged dependent variables:

$$y_i = \mathbf{x}_i \beta + \rho \mathbf{w}_{i\cdot} y_i + \epsilon_i.$$

If, however, we specify this differently, $\varepsilon_i = \lambda \mathbf{w}_{i\cdot} \xi_i + \epsilon_i$, we get

$$y_i = \mathbf{x}_i \beta + \lambda \mathbf{w}_{i\cdot} \xi_i + \epsilon_i,$$

which is a spatial error formulation.

We turn next to an examination of the spatially lagged dependent variable model; the spatial error model is addressed in Chapter 3.[2]

It is tempting to interpret the coefficient estimate for GDP per capita in the model with the spatially lagged *y* in Table 2.2 and compare this directly with Table 2.1, suggesting a seemingly larger effect of GDP per capita. However, this interpretation is not correct. The coefficient estimates have different interpretations, as the model with the spatially lagged *y* in Equation 2.1 is an autoregressive specification, so that the coefficient for the impact of *x* now reflects the short-run impact of x_i on y_i rather than the net effect, as is the case of the coefficient for *x* in the OLS model without the spatially lagged *y*. Since the value of y_i will influence the level of democracy in other states y_j and these y_j, in turn, feed back on to y_i, we need to take into account the additional effects that the short impact of x_i exerts on y_i through its impact on the level of democracy in other states.

This is analogous to the interpretation of the coefficient β for a covariate x_t in a time series model where we have a temporal lag of the dependent variable y_{t-1} on the right-hand side, for example,

$$y_t = \beta x_t + \phi y_{t-1} + \epsilon_t.$$

In this equation, β indicates the immediate effect of x_t on y_t. But this will, in turn, affect y_{t-1} in the following time period and the long-run effect of x_t must thus take into account the part of the net effect that works through the autoregressive part or the estimated coefficient for the impact of the lag y_{t-1}. The long-run effect of x_t will be $\beta/(1-\phi)$. In a situation where ϕ is large, the long-run effect $\beta/(1-\phi)$ can be substantially larger than β.

Continuing this analogy, imagine if we could increase the log of GDP per capita by one unit in only a single country i, which would have an immediate impact on that country's level of democracy of β_1. However, the model in Equation 2.1 implies spatial dynamics with a feedback effect between countries, where country i's level of democracy is also held to have an effect on its neighbors' level of democracy. Hence, an increase in democracy that affects i's level of democracy will then influence democracy in the neighbors of j. Contemporaneously, in turn, the neighbors' neighbors will also be affected, throughout all connected countries. In general, all countries will have some neighbors so that eventually the influence of all countries will be affected. But note that Equation 2.1 includes democracy for all countries in the system y, so if the democracy level of other countries connected to i increases so will the level of democracy in i. In this way, an exogenous shock to one observation, such as our thought experiment, will have a reverberating effect throughout the system with feedback among observations and flow through the system as a series of adjustments until it settles on some new stable equilibrium (Cressie, 1993; Lin, Wu, & Lee, 2006).

Rather than focusing on the coefficient estimates for x_i alone in a spatially lagged y model, it is important to consider the equilibrium effects. Unfortunately, the long-run effect for the spatially lagged y cannot be stated in a form as simple as in the case of the long-run effect in the presence of a temporally lagged y. We will return later for how to characterize and estimate the equilibrium effect of covariates in a spatially lagged y model. First, however, we turn to problems posed by the presence of the spatial lag of y on the right-hand side of the equation and the implied endogeneity problems related to consistent estimation of the model in the ordinary least squares setup.

The following section relies on matrix algebra and focuses on issues of estimation and why a maximum likelihood estimator (MLE) may be preferable to estimating the spatially lagged y model. Since using the MLE by itself does not require an understanding of all the details in this section, readers who are not interested in issues of estimation may skip the details in this section and proceed immediately to the next section.

2.2 Estimating the Spatially Lagged y Model

In a time series model with a temporally lagged y_{t-1} on the right-hand side, the presence of the temporal lag y_{t-1} does not create problems for estimation with OLS, provided there is no serial correlation in the residuals of the regression model. More precisely, OLS with a lagged dependent variable does not create problems for estimation, assuming the model is correctly specified. There has been considerable debate on the merits of including lagged dependent variables, but this is a debate on whether certain other assumptions of the data-generating process are reasonable. We refer to Keele and Kelly (2006) for a discussion. However, whereas y_{t-1} is predetermined at time t, the spatial lag of y is simultaneous and based on y itself. This simultaneity creates problems when estimating the spatially lagged y model. To understand why, it is helpful to look at the spatially lagged model in matrix algebra. Following the notation in Anselin (1988), the spatially lagged y model can be expressed as

$$
\begin{aligned}
Y &= \rho \mathbf{W}\mathbf{y} + \mathbf{X}\beta + \epsilon, \\
\epsilon &\sim N(0, \sigma^2 \mathbf{I}),
\end{aligned}
$$

where $\mathbf{I}$ represents the identity matrix (an $n \times n$ matrix with 1s on the diagonal and 0s everywhere else) and $\sim N(0, \sigma^2 \mathbf{I})$ indicates that the errors are distributed normally with a constant variance and that the cross-products of the error covariance matrix are 0. If $\rho = 0$, there is no spatial dependence, and the first part on the right-hand side cancels out, leaving us with the standard classical regression model where OLS is appropriate. However, if $\rho \neq 0$, we have simultaneity, and the OLS estimates will not converge to their "true" values as the number of observations increases. Instead, the feedback or dependency that is ignored by the OLS specification is likely to grow rather than be eliminated as the size of the data frame grows. Actually, it depends explicitly on the size and exact form of the connectivity matrix.

If using OLS to estimate the spatially lagged y model is problematic, what are the alternative estimation methods? The spatially lagged y model can be estimated using two-stage instrumental variable estimation, for example, using the exogenous variables $\mathbf{X}$, $\mathbf{WX}$, and $\mathbf{W}^2\mathbf{X}$ as instruments for the spatial lag of y. We do not cover estimation by instrumental variables in any detail here but instead focus on a suggested maximum likelihood estimator for the spatially lagged y model, which will be consistent and asymptotically efficient if the model is correctly specified. Although the OLS estimates of the spatially lagged y will face problems with simultaneity due to the presence of $\mathbf{Wy}$ on the right-hand side, the properties of the MLE hold only asymptotically, and the size of the inconsistency or bias will depend on the specific circumstances in each application. Franzese and Hayes (2007) explore the performance of different estimators through Monte Carlo simulation and suggest that the OLS estimates of the spatially lagged y model in some settings can have smaller mean squared errors than the MLE. Small samples make it difficult to leverage the spatial formulation.

Maximizing the likelihood for the spatially lagged y model is complicated. To see the complications, it is helpful to consider that the spatially lagged y model can be rewritten as follows:

$$\epsilon = y - \rho \mathbf{W} y - \mathbf{X}\beta = (\mathbf{I} - \rho \mathbf{W}) y - \mathbf{X}\beta.$$

This, in turn, implies that we can write the estimator for β:

$$\beta = (\mathbf{X}'\mathbf{X})^{-1}\mathbf{X}'(\mathbf{I} - \rho \mathbf{W}) y.$$

Finding the parameter estimates β for this model is difficult when ρ is unknown, as the log likelihood function involves the determinant $|\mathbf{I} - \rho \mathbf{W}|$. This is an nth-order polynomial in ρ, which must be evaluated at every iteration in the estimation procedure. However, Ord (1975) showed that if $\mathbf{W}$ has eigenvalues $(\omega_1, \ldots, \omega_n)$, then

$$|\omega \mathbf{I} - \rho \mathbf{W}| = \prod_{i=1}^{n} (\omega - \omega_i).$$

This, in turn, implies that

$$|\mathbf{I} - \rho \mathbf{W}| = \prod_{i=1}^{n} (1 - \rho \omega_i).$$

Ord suggested that the ω_i of $\mathbf{W}$ can be found at the outset, prior to the estimation of the rest of the model.

Recall that the log likelihood function for the classical linear regression model, assuming constant variances, is

$$\ln \mathcal{L}(\beta, \sigma^2) = -N/2 \ln(2\pi) - N/2 \ln(2\sigma^2) - (y - \mathbf{X}\beta)'(y - \mathbf{X}\beta)/2\sigma^2.$$

In contrast, the log likelihood function for the spatial lag model is

$$\ln \mathcal{L}(\beta, \sigma^2, \rho) = \ln|\mathbf{I} - \rho \mathbf{W}| - N/2 \ln(2\pi) - N/2 \ln(2\sigma^2) - (y - \rho \mathbf{W} - \mathbf{X}\beta)'(y - \rho \mathbf{W} - \mathbf{X}\beta)/2\sigma^2,$$

and treating ω_i as given prior to estimation, we can easily find the MLE for the spatially lagged y model by maximizing this function. We also need to ensure that the coefficients do not lead to explosive feedback processes, which would cause the covariance matrix to be nonpositive definite.

Although alternative algorithms are currently available, most implementations of the MLE for the spatially lagged y model still rely on Ord's insight to eliminate a major part of the computational complexity. In moving to the MLE, however, the basic assumptions change. In OLS, the errors are normally distributed, but the data need not be. In the MLE for spatial models, the data are assumed to be normally distributed.

2.3 Maximum Likelihood Estimates of the Spatially Lagged y Model of Democracy

In this section, we present maximum likelihood estimates for the spatially lagged y model of democracy and compare these with the OLS estimates of the same model.

We first show the code to implement this in R:

```
sldv.fit <- lagsarlm(democracy ~ log(gdp.2002/population),
   data=sldv, nb2listw(nblist), method="eigen", quiet=FALSE)
summary(sldv.fit)
moran.test(resid(sldv.fit),nb2listw(nblist))
```

This generates the results shown in Table 2.3. As one can see, this yields a higher estimate of the coefficient for GDP per capita (approximately 1.0) than the OLS estimates of the model (0.76) and a lower estimate of the $\hat{\rho}$ parameter for the spatial lag of y (0.56) than in the OLS estimates for the spatially lagged y model. Our key conclusions remain the same regardless of estimation methods, in the sense that including a spatially lagged y term notably improves the ability of the model to account for variation in democracy across countries.

TABLE 2.3
Maximum Likelihood Estimates of the Spatially Lagged y Model

	$\hat{\beta}$	$SE(\hat{\beta})$	*z Value*
Intercept	–6.20	2.08	–2.98
ln GDP per capita	0.99	0.28	3.59
$\hat{\rho}$	0.56	0.08	7.43
$N = 158$			
Log likelihood ($df = 4$) = –491.10			

If we believe that the MLE is generally more appropriate than OLS estimates for the spatially lagged y, we might conjecture that the OLS estimates underestimate the coefficient for GDP per capita and overestimate the coefficient for the spatial lag. This conjecture is not testable, since we do not know what the "true" parameters might be and how closely our model resembles these or whether there are "true" parameter values that even exist.

The Lagrange multiplier test for residual autocorrelation is the preferred test on residuals from a spatial model. The test has a value of 2.1 with an associated probability of 0.147 in this example, allowing a clear rejection of a remaining first-order autoregression among the residuals. By comparison, the estimated Moran's $\mathcal{I}$ for residual spatial clustering produces the same result, as does the saddlepoint modification. The former has a standard score –0.46, which also points toward a rejection of the notion of simple spatial correlation among the residuals based on the same connectivity matrix **W**. If we examine the residuals for remaining spatial patterns for the OLS with the spatially lagged y, we find considerable evidence that the residuals still exhibit strong spatial clustering, with a Moran $\mathcal{I}$ of –0.17 and associated standard score of –3.21; saddlepoint estimates are virtually identical. A negative Moran $\mathcal{I}$ suggests a repulsive pattern among the residuals, lending some support to our conjecture that the OLS estimates overestimate the impact of the spatially lagged y and in this sense overcorrect for spatial dependence in assessing the effects of GDP per capita. We emphasize that these autocorrelation tests of residuals should be used cautiously because they are dependent on the connectivity matrix, which itself is subject to a variety of plausible specifications in most instances. We return to this point below.

2.4 Equilibrium Effects in the Spatially Lagged y Model

With the maximum likelihood estimates for the spatially lagged y model in hand, we explore the equilibrium effects of GDP per capita on democracy. This requires taking into account the implications a change in an independent variable in one state i will have on other states. This leads through the connectivity matrix to a type of chain reaction in other states that would eventually return to influence y_i via the spatially lagged y term.

Keep in mind that the spatially lagged regression model can be written in matrix notation as

$$y = \mathbf{X}\beta + \rho\mathbf{W}y + \epsilon.$$

Moving all terms that involve the dependent variable y to the left-hand side, we get

$$(\mathbf{I} - \rho\mathbf{W})y = \mathbf{X}\beta + \epsilon.$$

Solving this equation for y, and taking expectations, we then find that, in equilibrium, the expected value for y will be

$$E(y) = (\mathbf{I} - \rho\mathbf{W})^{-1}\mathbf{X}\beta.$$

It is obvious that $E(y)$ will reduce to $\mathbf{X}\beta$ only if $\rho = 0$. To determine the expected value of y_i or the equilibrium effect of x, we must consider the spatial multiplier $(\mathbf{I} - \rho\mathbf{W})^{-1}$. This multiplier tells us how much of the change in x_i will "spill over" onto other states j and in turn affect y_i through the impact of y in the spatial lag. This is similar to the Leontief (1986) inverse used in input-output analysis to evaluate how change in demand in one sector will influence total production in a multisectoral system.

Hence, to determine the equilibrium impact of a one unit difference for some observation in x_i we need to premultiply a vector $\Delta x(i)$, where the value for other units j is held constant, by $(\mathbf{I} - \rho\mathbf{W})^{-1}\beta$. Since all states will have different degrees of connectivities to other states and different degrees of higher-order connectivities with others, it follows that the impact of a given change in x_i will depend on the particular country changed. Imagine a situation where we have two disjoint regions without any bridging ties. In such a case, a change in Region 1 would affect other countries in Region 1, but these changes would have no effect on countries in Region 2.

A useful way to illustrate the variation in equilibrium effects is to consider the impact of a change for all different countries and examine the distribution of the country-specific estimates. In the example shown here, we have a mean equilibrium effect of 1.09, which is about 10% higher than the short-run impact of the log of GDP per capita given by the coefficient estimate $\hat{\beta}$ of 0.99 in Table 2.3. The individual country-specific equilibrium effects range from a low of 1.03 (for Mongolia) and a high of 1.24 (for Papua New Guinea), which is about 25% higher than the short-run effect for GDP per capita. Clearly, we should be hesitant to make inferences about the effects of a covariate x in a spatially lagged y model without considering the spatial multiplier and the variation that will exist across spatial units. Figure 2.1 displays a histogram of the estimated effects.

To understand how one state's GDP per capita affects the expected value of democracy in other states, it is instructive to examine the full vector $(\mathbf{I} - \rho\mathbf{W})^{-1}\beta\Delta x(i)$. We use Russia as an illustration. Table 2.4

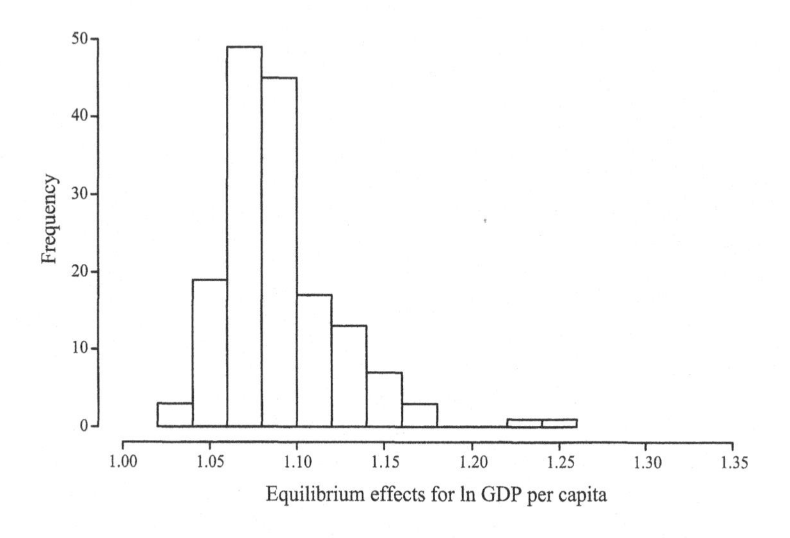

Figure 2.1 Histogram of the Equilibrium Effects for ln GDP Per Capita.

displays the 10 highest values of $(\mathbf{I}-\rho\mathbf{W})^{-1}\,\beta\Delta x(i)$ based on Russia, the estimates for the spatially lagged y model reported in Table 2.3, and the connectivities specified in **W**. As we see, the implied equilibrium impact for Russia would be 1.09, which is close to the median of the equilibrium impacts implied by the model. The values for the other states indicate that a change in Russia would have implications for Russia's neighbors in Asia and Europe. To see what these estimates imply in substantive terms, recall that the coefficients for the estimated impact pertain to the log of GDP per capita. A 10% change in the current GDP per capita of Russia (i.e., $2,279) would only raise its own predicted value of democracy by a little over 0.1 points on the POLITY scale. For the country with the largest equilibrium impact of change in the GDP per capita value for Russia, the increase in the predicted value of democracy would be only a little over 0.02 based on these estimates. This reinforces our conclusion that even very large differences in the GDP per capita of one state would not change the expected level of democracy around the world very much according to this model, and the impact of ln GDP per capita is substantially lower in the spatially lagged y model taking into account the influence of the level of democracy in connected states than the OLS results where we treat the individual observations as independent of one another.

TABLE 2.4
Equilibrium Impacts of log GDP Per Capita
for Russia, 10 Highest Values

Country	*Impact*
Russia	1.09
People's Republic of Korea	0.24
Japan	0.24
Mongolia	0.24
Finland	0.22
Estonia	0.21
Norway	0.20
Lithuania	0.20
Latvia	0.12
Armenia	0.18

We show below the code to construct such an experiment in abbreviated form, based on the estimates for the spatially lagged *y* object shown above:

```
# Code to calculate equilibrium effect of changes in GDP per capita
# Create vector to store the estimate for each state
ee.est <- rep(NA,dim(sldv)[1])
# Assign the country name labels
names(ee.est) <- sldv$tla
# Create a null vector to use in loop
svec <- rep(0,dim(sldv)[1])
# Create a N x N identity matrix
eye <- matrix(0,nrow=dim(sldv)[1],ncol=dim(sldv)[1])
diag(eye) <- 1
# Loop over 1:n states and store effect of change in
# each state i in ee.est[i]
for(i in 1:length(ee.est)){
     cvec <- svec
     cvec[i] <- 1
     res <- solve(eye - 0.56315 * wmat) %*% cvec * 0.99877
     ee.est[i] <- res[i]
}
# Russia example of impact on other states (observation 120)
cvec <- rep(0,dim(sldv)[1])
cvec[120] <- 1
# Store estimates for impact of change in Russia in rus.est
eye <- matrix(0,nrow=dim(sldv)[1],ncol=dim(sldv)[1])
```

```
diag(eye) <- 1
rus.est <- solve(eye -  0.56315 * wmat) %*% cvec*0.99877
# Find ten highest values of rus.est vector
rus.est <- round(rus.est,3)
rus.est <- data.frame(sldv$tla,rus.est)
rus.est[rev(order(rus.est$rus.est)),][1:10,]
```

The previous results from the OLS model suggested that GDP per capita had a relatively limited effect on expected level of democracy. The MLE results for the spatially lagged y model in Table 2.3 also suggest a relatively small immediate impact of GDP per capita. When we explored the long-run equilibrium effects of changes in GDP per capita we found a somewhat larger but still quite limited impact. What do the coefficients in Table 2.3 imply about the relationship between a country's expected democracy level and that of its neighbors? Figure 2.2 shows graphically the expected covariation implied by the model. In this figure, we plot the expected value of the dependent variable (democracy, y) as a function of the level of democracy in neighboring states (the spatial lag, y^s and the independent variable, logged GDP per capita. It is clear from this contour mapping that the impact of GDP per capita is weak, but the spatial component has a strong impact on a state's expected level of democracy.

The expected level of democracy suggested by the model varies dramatically for a country, conditional on a constant GDP per capita. A country at the median GDP per capita with consistently autocratic neighbors (i.e., $y_i^s = -10$) would have an expected democracy score of about –4, while its expected level of democracy would be almost 7 if all its neighbors are democracies (i.e., $y_i^s = 10$). As such, although GDP per capita appears to have a relatively limited ability in accounting for variation in democracy in these empirical results, there is a very close relationship between a country's level of democracy and that of its neighbors.

Another way to think of these results is in terms of what would happen if democracy changed due to features not in the systematic part of the model (e.g., a shock in y_i for some country i) along with the short-term impact that this would have on the predicted level of democracy of other states j, $\hat{y}_j$ implied by the model. For example, consider what would happen if China were to become a democracy (i.e., a POLITY score of 10), as opposed to its current value of –7. What would the implications be for other states, using the estimated spatial model as our guide? Obviously, this will depend on not just $\hat{\rho}$ but also on the structure of the connectivity matrix **W** used in constructing the spatial lag. Recall that since **W** is row standardized, a change in China will carry less weight for countries

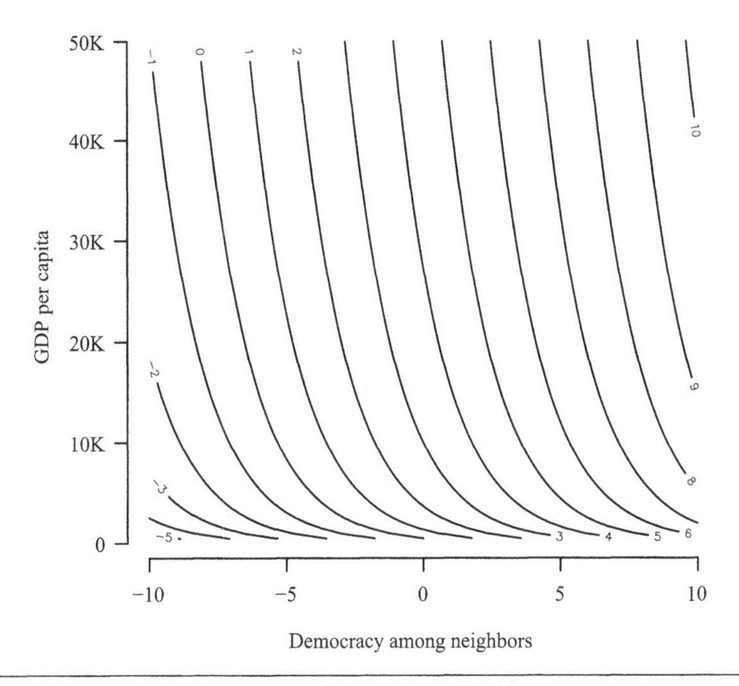

Figure 2.2 Plot of Expected Democracy by Variation of Mean Democracy Level Among Neighbors and the Logged GDP Per Capita.

that have many connected observations than countries with few neighbors. Table 2.5 displays the immediate impact on its 18 neighbors if China had a democracy score of 10. Taiwan, North Korea, and Mongolia, which each have China as one of their three neighbors in **W**, would be expected to see an increase in their democracy score of almost two points, while the immediate impact for Russia, with 20 neighboring states, would be only 15% of the direct impact for the states with China as one of their three neighbors.

We reproduce below the code to examine the impact of a change in y for China on other states in the system, based on the estimates for the spatially lagged y object shown above:

```
# Impact of change in $y$ to 10 in China
# China is observation 32
cvec <- rep(0,dim(sldv)[1])
cvec[32] <- 10
```

```
# Store estimates of change in China in chn.est
chn.est <- c(cbind (0, 0, wmat%*%cvec) %*%
     c(summary(sldv.fit)$Coef[,1],summary(sldv.fit)$rho))
chn.est <- round(chn.est,3)
# Find all states where non-zero impact
chn.est <- data.frame(sldv$tla,chn.est)
chn.est <- chn.est[rev(order(chn.est$chn.est)),]
chn.est[chn.est$chn.est>0,]
```

TABLE 2.5
Effects on Predicted Democracy $\hat{y}$ If China Had a POLITY Score of 10

Country	*Impact*
Taiwan	1.88
North Korea	1.88
Mongolia	1.88
Nepal	1.41
Bhutan	1.41
Pakistan	1.13
Laos	1.13
Kyrgyzstan	1.13
Bangladesh	1.13
Uzbekistan	0.94
Thailand	0.94
Myanmar/Burma	0.94
Tajikistan	0.80
India	0.80
Vietnam	0.80
Afghanistan	0.80
Kazakhstan	0.70
Russia	0.28

2.5 Spatial Dependence in Turnout in Italy

Shin (2001) and Shin and Agnew (2002, 2007a, 2007b) have studied the geographical distribution of political activity in Italy over the past several decades and have suggested important spatial dynamics in turnout and voting outcomes. We use their data to explore a simple version of the idea that the spatial variation in turnout rates can be accounted for by the geographic distribution of wealth and income in Italy. We use data from the Italian National Elections in 2001 and data on GDP per capita taken from each province in 1997. These are available for each of the 477 *collegi*, or single member districts (SMDs hereafter), that existed during this election.

2.5.1 Maps of the Main Variables

We illustrate the maps of the geographical distribution of voting turnout and GDP per capita as Step 1 in our spatial analysis (see Figures 2.3 and 2.4).

Turnout is highest in the north, especially in the far north, around Milan, and in Emilia-Romagna and Tuscany. Rome and Venice also have high turnout rates. In Modena, for example, turnout is more than 90%. In contrast, turnout hovers in the mid-teens in Sicily; even in the outskirts of Naples, voting turnout is only about 60%. The richest part of Italy, in terms of GDP per capita, is Lombardy. Income in the richest northern SMDs is

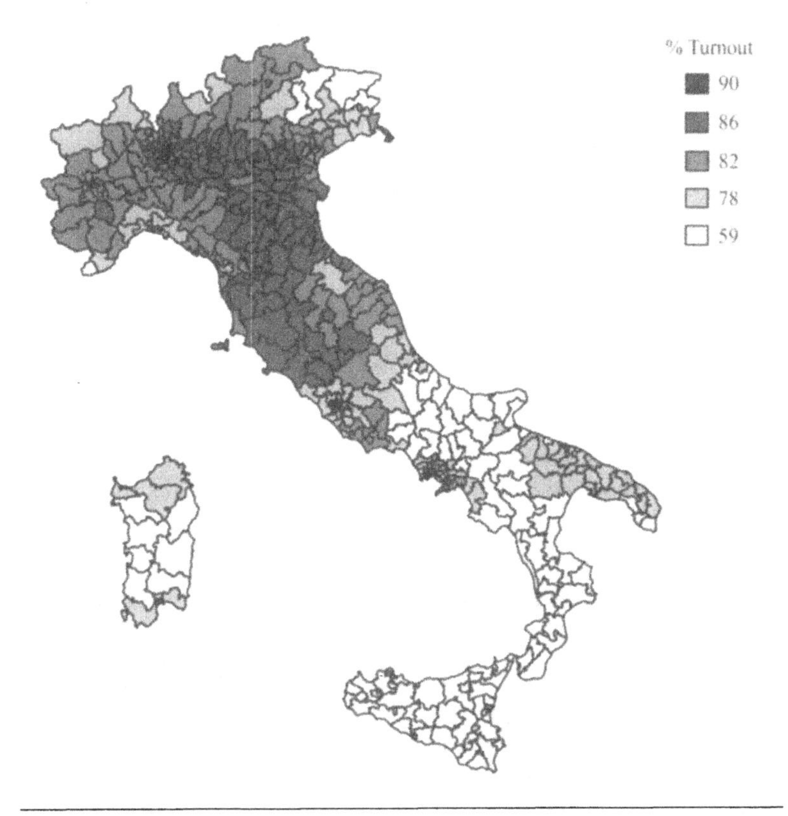

Figure 2.3 Voting Turnout by Collegio in Italy.

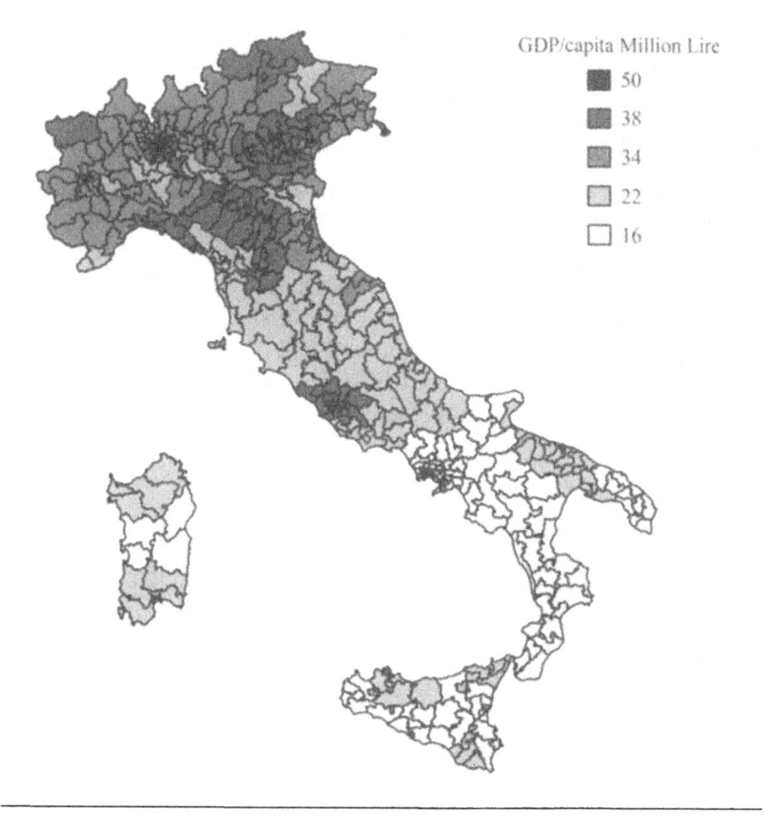

Figure 2.4 Per Capita GDP in Italy, Taken From 1997 Data.

about 1.5 times as large as per capita income in the poorest SMDs in the South. Clear clustering of both turnout and GDP per capita is apparent in this set of heuristic maps.

2.5.2 Calculation of Moran's $\mathcal{I}$ Statistics

In this section, we assess spatial clustering in turnout and GDP per capita more formally using the Moran $\mathcal{I}$ statistic. As a first cut at the spatial connectivities in Italy, we simply used the nearest neighbor distances for 50 km. We calculated the centroids of each district and then determined whether it was joined to the centroids of other districts by a distance of 50 km or less. By way of summary, we found that two Milan districts were connected to 54 others: ten and six. Eight SMDs were connected only to a single other district, but this mainly had to do with edge effects: An example is Trentino-Alto Adige, hidden in the Alps near the Brenner pass and Austria. On average, however, SMDs are linked to about 17 other districts by this 50-km rule.

Summaries can be easily obtained in R by a summary of the neighborhood object. As an example, we created these linkages by the following code:

```
tr <- readShapePoly("turnout",
  IDvar="FID_1", proj4string=CRS("+proj=robin +lon 0=0"))
dnn50km <- dnearneigh(coordinates(tr), 0, 50000)
summarize(dnn50km)
```

There are two types of Moran's calculations, one done under the assumption of randomization and the other done using a normality assumption. Regardless of whether we assume randomization or normality, Moran's $\mathcal{I}$ statistic indicates strong spatial patterning in these data. In these two tests, we find a Moran's $\mathcal{I}$ of 0.86 for both assumptions; turnout has similarly high values, 0.79 (for both). All these values are unusual, and they suggest a strong spatial patterning of both GDP per capita and turnout.

2.5.3 Regression Analysis

Turnout is clearly likely to be associated with differences in GDP per capita, but can the spatial clustering in turnout be accounted for by geographical variation in GDP alone? The simple model examined here is that turnout is a function of GDP per capita. This is first examined by a standard least squares estimation, which is presented in Table 2.6. The standard

TABLE 2.6
OLS Regression of Voting Turnout on GDP Per Capita, Logged in Italy in 1997

	$\hat{\beta}$	$SE(\hat{\beta})$	*t Value*
Intercept	35.30	2.21	15.96
ln GDP per capita	13.46	0.65	20.84

$N = 477$
Log likelihood ($df = 3$) = −1387.57
$F = 434.4$ ($df_1\_ = 1$, $df_2\_2 = 475$)

results indicate that income is a strong predictor of voting patterns in Italy and that a unit difference in GDP per capita (in millions of Lire) is associated with about 14% voting turnout. However, the Moran test for simple spatial patterns of the residuals has a value of 0.47, indicating that there is a spatial pattern not accounted for by the covariates.

We then examined a spatially lagged *y* regression model with the following code:

```
shin <- read.csv("italyturnout.csv",sep=",",header=T)
sldv.fit <- lagsarlm(turnout ~ log(gdpcap), data=shin,
      nb2listw(dnn50km), method="eigen", quiet=FALSE)

summary(sldv.fit)
```

The results are illustrated in Table 2.7. The effects of GDP per capita on turnout are less "strong" than the OLS results shown above, but more plausible. They suggest a smaller impact of income, but a strong impact nonetheless. However, the spatial lag variable has considerable salience.

TABLE 2.7
Spatially Lagged Regression of Voting Turnout on GDP Per Capita, Logged in Italy in 1997

	$\hat{\beta}$	$SE(\hat{\beta})$	*z Value*
Intercept	4.70	1.66	2.80
ln GDP per capita	1.77	0.48	3.66
$\hat{\rho}$	0.87	0.02	36.7
$N = 477$			
Log likelihood ($df = 3$) = –1,193			

2.5.4 Equilibrium Analysis

Following the approach specified above, it is simple to calculate the equilibrium values for each of the 477 SMDs, that is, the *expected values* given the model. We do not present these here but instead conduct a simple experiment in which we hypothetically double the GDP per capita of one of the poorest areas in Italy—Reggio Calabria-Sbarre. In so doing,

Figure 2.5 Increases in Expected Turnout as a Function of a Doubling of GDP Per Capita in a Single, Poor Collegio (Regio Calabria-Sbarre) in the South of Italy.

we calculate the difference in the expected value under this "scenario" versus the expected value given the model and the observed data. The differences for most SMDs are nonexistent, but in 15 other SMDs, there are differences in expected voting turnout of 1% or more as a result of this hypothetical change in the GDP per capita of a single SMD. As expected, the biggest changes are in neighboring SMDs. Figure 2.5 shows the resulting distribution of implied differences in turnout in Italy.

We show below the code to construct this experiment (in abbreviated form) for Reggio Calabria-Sbarre (observation 432) based on the spatially lagged y object previously created:

```
# Extract estimated rho
rho <- coef(sldv.fit)[3]

# Extract estimated beta
beta <- coef(sldv.fit)[1:2]

# Create a X matrix
X <- cbind(1,log(shin$gdpcap))

# Create an alternative X matrix, changing value for
# Reggio Calabria-Sbarre (obs 432)
Xs <- X
Xs[432]<- log(35)

# Create an identity matrix
I <- diag(length(shin$gdpcap))

# wmat is the weights matrix
wmat <- nb2mat(dnn50km,style="W")

# Find equilibrium effect by looking at
# the difference in expected value for the
# two X matrices
Ey <- solve(I - rho*wmat)%*%(X%*%beta)
EyS <- solve(I - rho*wmat)%*%(Xs%*%beta)
dif <- EyS-Ey
```

2.6 Using Different Weights Matrices in a Spatially Lagged Dependent Variable Model

We illustrate the impact of spatial weights matrices with reference to data on the U.S. Presidential Election in 2004.[3] These are easily extracted to an XML table and converted to a csv file. To simplify our task, we do not consider the cases of Alaska and Hawaii, as these are sufficiently distant from all other states to create some challenges in the analysis of regional data. The major variable we are interested in is the share of the total vote received by George W. Bush and John F. Kerry in each of the 48 contiguous states, plus the District of Columbia. For the purposes of this exercise, we ignore the write-in votes from each state. We construct a ratio of the Bush votes to the Kerry votes and use this as the dependent variable.

To answer the question of how much autocorrelation there might be among these data in the context of regional patterns, we create several

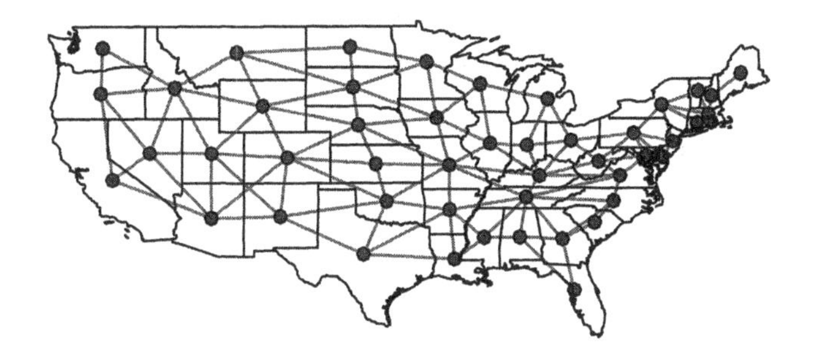

Figure 2.6 Map of 48 U.S. States and Their First-Order Connectivities, Based on Shared Borders Using Contiguity.

measures of the spatial connectivities among these 49 political and geographical units. The first such measure is simply a measure of the contiguity of states. In this context, Washington state is neighbored by Idaho and Oregon, because they share borders. Colorado shares borders with New Mexico, Arizona, Utah, Wyoming, Nebraska, Kansas, and Oklahoma. At the other end of the spectrum, Maine has only one border. These are illustrated in Figure 2.6.

The sample code to construct this map is provided below:

```
library(maptools);library(network)
library(spdep);library(sp);library(rgdal)
setwd("...")

# read in 2004 presidential votes
presvote <- read.table("2004presvote.csv",sep=",",header=T)

# read in shape files for 48 US States plus District of Columbia
# will create a MAP OBJECT
# use equal area projection (Robinson)
usa.shp <- read.shape("48_states.shp")

usaall <- merge(usa.shp$att.data, presvote,
      by.x = "STATE_NAME", by.y = "State",
      sort = F)

# Create a distance matrix from original polygon shape file
tr <- readShapePoly("48_states.shp",
  IDvar="ObjectID", proj4string=CRS("+proj=robin +lon 0=0"))
centroids <- coordinates(tr)
```

```
# Create polygons in a spatial object
us48polys <- Map2poly(usa.shp,
      region.id = as.character(usa.shp$att.data$STATE_NAME))

# Create neighbors, list, and matrix objects from polygon
      centroids
us48.nb <- poly2nb(us48polys,
      row.names = as.character(usa.shp$att.data$STATE_NAME))
us48.listw <- nb2listw(us48.nb, style = "B")
us48.mat <- (nb2mat(us48.nb,style="B"))

# plot the network among the centroids
colnames(us48.mat) <- rownames(us48.mat) <- usa.shp$att.dat$STATE_ABBR
usa <- network(us48.mat,directed=F)

set.seed(123)
# plot network first; then add state boundaries
plot.network(usa1,displayisolates=T,displaylabels=F,
      boxed.labels=F,coord=centroids,label.col="gray20",
      usearrows=F,edge.col=rep("gray60",190),
      vertex.col="gray30",edge.lty=1)
plot(us48polys,bty="n",border="slategray3",forcefill=TRUE,
      xaxt="n",yaxt="n",lwd=.000000000125,las=1,
      ylab="",xlab="",add=T)
```

Next we turn to mapping of the ratio of votes received by George W. Bush and John F. Kerry in each state during the 2004 Presidential Election. As shown in Figure 2.7, there appears to be a strong geographical patterning of the votes by state during the 2004 Presidential Elections.

```
library(RColorBrewer)
# now plot the Bush:Kerry vote ratio
bk <- usaall$Bush/usaall$Kerry
# set up five categories and assign colors
breaks <- round(quantile((bk), seq(0,1,1/5), na.rm=TRUE),1)
cols <- brewer.pal(length(breaks), "Greys")

# use findInterval to color states by bk variable
plot(us48polys,bty="n",border="slategray3",forcefill=TRUE,
      xaxt="n",yaxt="n",lwd=.000000000125,las=1,ylab="",xlab="")
plot(us48polys,bty="n",col=cols[findInterval(bk,breaks,
      all.inside=T)],forcefill=T,add=T)
legend(x = c(-125, -115), y = c(27, 32), legend = leglabs(breaks),
      fill = cols, bty = "n")
```

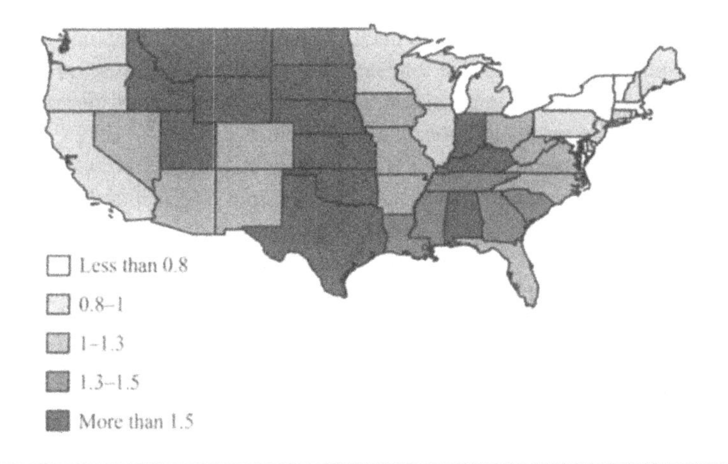

Figure 2.7 Ratio of Bush Votes to Kerry Votes in Each State During the 2004 Presidential Election.

Moran's $\mathcal{I}$ also shows numerical evidence of this patterning (see Table 2.8).

The average gross state product (akin to GDP) is measured by the Bureau of Economic Analysis, part of the U.S. Department of Commerce. The most recent annual data are available at http://www.bea.gov/bea/newsrelarchive/2006/gsp1006.xls, which also includes the growth rate of gross state product from 1997 to 2004. These data characterize the vibrancy of the state economy in the 7 years prior to the 2004 election. We use these data as a covariate to explain votes in the 2004 Presidential Election.

TABLE 2.8
Autocorrelation in the Bush-Kerry Vote Totals During the 2004 Presidential Election

No.	*Standard Score*	*Weights Schema*
Moran's $\mathcal{I}$		
0.39	4.7	Bordering states
0.49	5.7	Nearest 4 neighbors
0.30	7.0	Nearest 12 neighbors
Geary's $\mathcal{C}$		
0.65	–2.7	Bordering states
0.65	–3.6	Nearest 4 neighbors
0.69	–5.1	Nearest 12 neighbors

We set up two basic spatial connectivity matrices, one based on contiguity and the other based on the four nearest neighbors. A regression with a spatially lagged dependent variable was estimated for each of the contiguity codings. The results are presented in Table 2.9.

The empirical results suggest that states with a higher growth rate in GDP will have relatively fewer votes for George W. Bush and relatively more votes for John F. Kerry. For the case of contiguity coding, there is evidence of a weak, positive spatial correlation (0.09), while the spatial weight encoding that uses the four nearest neighbors shows a stronger level of positive correlation in the Bush-Kerry vote ratio (0.60). These two different estimations not only produce different results in the standard regression output tables, but more importantly will result in equilibrium values that are substantially different. As shown in Figure 2.8, the distributions of equilibrium effects for these two different weighting schemes are quite distinct, even though they are positively correlated with one another. The simple contiguity of borders has a mean effect in equilibrium that is more modest (–0.15) than is obtained using the nearest four neighbors (–0.35). The bottom line of this simple example is that the weights matrix is a very important aspect of spatial analysis and even relatively small perturbations in the weights matrix will have salient consequences in the empirical results.

TABLE 2.9
Spatially Lagged Regression of Ratio of Bush to Kerry Votes by State in the 2004 U.S. Presidential Elections on GDP Growth Rates (1997–2004) in Each State

No.	$\hat{\beta}$	$SE(\hat{\beta})$	*z Value*
Queen contiguity			
Intercept	0.86	0.21	4.00
GDP growth rate	–0.05	0.06	0.85
$\hat{\rho}$	0.09	0.02	20.4
$N = 49$			
Log likelihood ($df = 3$) = –25.63			
Four nearest neighbors			
Intercept	0.63	0.23	2.72
GDP growth rate	–0.06	0.05	1.04
$\hat{\rho}$	0.60	0.12	18.4
$N = 49$			
Log likelihood ($df = 3$) = –25.19			

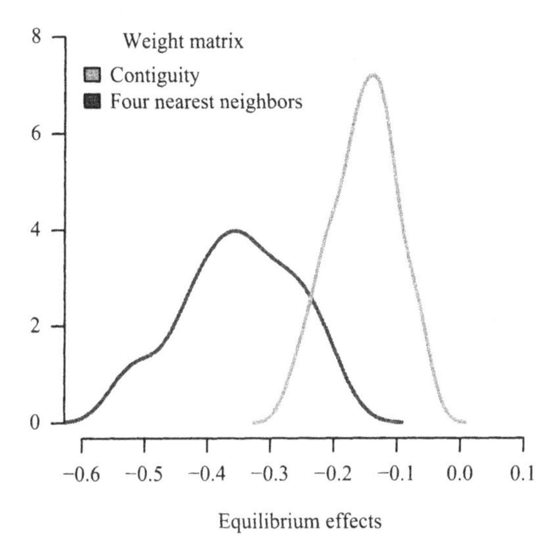

Figure 2.8 Posterior Distribution of Equilibrium Effects of Spatially Lagged Dependent Variable With Different Weighting Schemas.

2.7 The Spatially Lagged Dependent Variable Versus OLS With Dummy Variables

Social scientists often recognize that there is considerable heterogeneity between different regions of the world and that the country-specific covariates included in their regression model are unable to account adequately for this spatial variation. A common way to address spatial heterogeneity is to include dummy variables for different geographical regions. Such dummy variables will essentially fit separate intercepts for different geographical regions, thereby allowing for taking into account fixed, mean differences in the dependent variable y across discrete regions. This is by far the most common approach for addressing regional heterogeneity in applied work, and social science is replete with models in which "region" categories are included as dummy variables. Moreover, such models are becoming more common as analysts are increasingly concerned that pooled OLS estimates may fail to take into account important region-specific differences.

For example, Lee (2005), in a study examining the impact of democracy and the size of the public sector on income inequality, fits regional dummy variables for Africa, Asia, and Latin America and argues that the latter two regions appear to differ notably from the reference category (Organization for Economic Cooperation and Development) over and beyond what can be accounted for by other country-specific right-hand side variables in the model. In the context of studies of democracy, Burkhart and Lewis-Beck (1994) estimate models by considering heterogeneity in levels of democracy across different world system positions through dummy variables, distinguishing between countries in the core, periphery, and semiperiphery of the world economy.

Models with regional dummy variables are clearly popular in the social sciences and provide an alternative to the spatially lagged y model. We first present an alternative of the original OLS model, adding regional dummy variables, and then comment on the relationship between this model and the spatially lagged y model. Table 2.10 presents a model, with dummy variables for countries in Latin America and the Caribbean, Europe, Sub-Saharan Africa, the Middle East and North Africa, Asia, and Oceania. The omitted region, or reference category, is North America (i.e., the United States and Canada). The coefficient estimate for the different regions indicates the predicted difference in level of democracy, holding GDP per capita constant, for a country in a given region relative to North America. Latin America and the Caribbean, Europe, and Oceania seem to have levels of democracy on average essentially indistinguishable from North America, while Sub-Saharan Africa and Asia and, in particular, the Middle East and North Africa tend to have much lower average levels of democracy. We also note that the coefficient estimate for the log GDP per capita is much lower here than in the case of the OLS model treating all

TABLE 2.10
Estimates of Model With Regional Dummy Variables

No.	$\hat{\beta}$	$SE(\hat{\beta})$	*t Value*
Intercept	−1.89	5.06	−0.37
ln GDP per capita	1.15	0.34	3.39
Latin America and the Caribbean	0.09	3.84	0.02
Europe	−0.41	3.74	−0.11
Sub-Saharan Africa	−4.71	3.97	−1.19
The Middle East and North Africa	−11.77	3.85	−3.05
Asia	−5.97	3.92	−1.52
Oceania	0.90	4.72	0.19

$N = 158$
Log likelihood ($df = 8$) = −477.52
$F = 18.65$ ($df_1 = 7$, $df_2 = 150$)

countries as independent of one another (i.e., 1.68). In fact, the coefficient estimate for the log of GDP per capita in this model is quite similar to the average of the equilibrium impact found for the spatially lagged y model discussed above (i.e., 1.09). This lends some support to the argument that the pooled OLS disregards a great deal of regional heterogeneity and that controlling for differences across regions through fitting dummy variables helps address the implications of regional heterogeneity, as well as the implications for overestimating the impact of GDP per capita.

Is the regional dummy approach a suitable alternative to the spatially lagged y model? One possible way to answer this question would be to look at the parsimony of the two models. Although the OLS with dummy variables has a somewhat higher log likelihood, it achieves this by fitting six new parameters, or five more than the spatially lagged y model. Moreover, the regional dummy variable model by itself does not contain a generative story of how these regional differences originate but simply fits separate intercepts based on the observed variation across regions. If one of the members of a region were to see a change in its level of democracy as a result of a change in its GDP per capita, the predicted values of the other countries would not change since the regional differences are treated as fixed and countries do not influence one another. In contrast, the spatially lagged y model fits only one additional parameter, which can be interpreted substantively as the impact of the level of democracy y among connected countries on a country's level of democracy. If one is not concerned about parsimony, but simply wants to maximize fit, it would obviously be possible to fit a spatially lagged y model with regional dummy variables. In this case, this model still suggests residual spatial clustering, in the sense of returning a statistically significant and positive estimate of $\hat{\rho} = 0.25$. Fitting a spatially lagged y model with regional dummy variables requires the analysts to reconcile the assumption of fixed regional differences with the implied endogeneity of the spatially lagged y specification. In many instances, it may be difficult to estimate separate parameters for regional dummies and the spatially lagged y, if the connectivities in $\mathbf{W}$ are very "similar" to the regional demarcations, in ways similar to the problem of collinear regressors. Furthermore, note that the regional dummies by construction assign countries to discrete, or proper name regions, whereas the spatially lagged y term in this case is based on a connectivity matrix $\mathbf{W}$ where the connectivities are specific to each country. Compared with the discrete specification, the country-based connectivity specifications have the advantage of not forcing countries quite far apart geographically such as Greece and Ireland to belong to the same cluster and not forcing countries that span several commonly defined geographical regions such as Turkey and Russia to belong to one region only.

Even if we believe that mutually exclusive regions are appropriate to specify connectivities between observations, the regional dummy variable specification may not generally be an adequate alternative to the spatially lagged y model and entails additional assumptions that we find overly restrictive. To see this, consider a regression with k different dummy variables D_k where $y = b_1 D_1 + \cdots + b_k D_k + e$. Unlike the connectivity list or matrix where i is not included as a neighbor of itself, each region here includes both i and all its neighbors. But if the number of cases in each of the regions is large, then it is possible to show that $\mathbf{W}y \approx b_1 D_1 + \cdots + b_k D_k$. This means that the dummy variable regression can be rewritten as $y = b_1 D_1 + \cdots + b_k D_k \approx \mathbf{W}y + e$. In this sense, the dummy variable regression model is a special case of the spatially lagged y model, which simply assumes a $\rho = 1$ rather than estimating this parameter empirically (Lin et al., 2006). In other words, the regional dummy variable assumes that all observations within every region are homogeneous and interconnected, whereas the spatially lagged y model allows the degree of similarity to be estimated. Moreover, the spatially lagged y model can easily handle cases with a wide range of forms of connectivity, while the dummy variable approach assumes disjoint clusters where every unit of analysis within a given cluster is connected to everyone else and there are no links between clusters; nor can units belong to a variety of different clusters.

Notes

1. Using the saddlepoint adjustment based on the Barndorff-Nielsen approach yields a slightly reduced assessment: the associated Z score is 6.9.

2. Further such *decompositions* can also be envisioned, such as the introduction of hierarchical covariates that selectively apply to observations in specific regions or administrative districts. Such models are not addressed in this book.

3. The 2004 election data are available at http://www.fec.gov/pubrec/fe2004/federalelections2004.pdf.

3. SPATIAL ERROR MODEL

In Chapter 2, we examined the spatially lagged dependent variable model in which "neighboring" values of the dependent variable exert a direct effect on the value of the dependent variable itself. Although this is probably the most common, and perhaps the most generally useful way to think about spatial dependence, it is not the only possible way to represent spatial dependence in a linear model with a continuous dependent variable. In this chapter, we examine an alternative conception in which the spatial dependence enters through the errors rather than through the systematic component of the model. Such a model is typically called the *spatial error model.* We also examine an important possible extension of spatial regression models to distance concepts based on metrics other than geography in the context of the spatial error model.

3.1 The Spatial Error Model

Whereas the spatially lagged dependent variable model sees spatial dependence as substance, in the sense that the y_i is influenced by the value y_j for other countries $(j \neq i)$, the spatial error model treats spatial correlation primarily as a nuisance, much like how statistical approaches often treat temporal serial correlation as something to be eliminated and solely as an estimation problem. This approach generally focuses on estimating the parameters for the independent variables of interest in the systematic part of the model and essentially disregards the possibility that the observed correlation may reflect something meaningful about the data generation process. Instead of letting y_i affect y_j directly, the spatial error model assumes that the errors of a model are spatially correlated. There are myriad ways in which this could be specified. We focus on a simple one that is based on a coding of the spatial regime in terms of spatial weights; other important approaches focus on geostatistical covariance structures, but we do not examine these in this book. Following the earlier notation, if we let $\mathbf{w}_i$ denote the vector of $\mathbf{W}$ indicating how close other units $j \neq i$ are to i, we can write the spatial error model as follows:

$$y_i = \mathbf{x}_i\beta + \lambda \mathbf{w}_{i\cdot}\xi_i + \epsilon_i.$$

Here we have decomposed the overall error into two components—namely, ϵ, a spatially uncorrelated error term that satisfied the normal regression assumption, and ξ, which is a term indicating the spatial component of the error term. The parameter λ indicates the extent to which the spatial component of the errors ξ are correlated with one another for nearby observations, as given by the vector of connectivities $\mathbf{w}_i$. Alternatively, we can state the spatial error model in matrix form, based on the terms previously defined in Chapter 2:

$$\begin{aligned} y &= \mathbf{X}\beta + \lambda\mathbf{W}\xi + \epsilon, \\ \epsilon &\sim N(0, \sigma^2\mathbf{I}). \end{aligned}$$

If there is no spatial correlation between the errors for connected observations *i* and *j*, the spatial error parameter λ will be 0, and the model reduces to the standard linear regression model where the individual observations are independent of one another and we can proceed to estimate the model by OLS in the conventional manner. However, if the spatial error parameter $\lambda \neq 0$, then we have a pattern of spatial dependence between the errors for connected observations. This could simply be coincidental, or it could reflect other kinds of mis-specifications in the systematic component of the model, in particular omitted variables that are spatially clustered. Social scientists typically expect to see positive spatial correlation. This implies the clustering of similar values; that is, the errors for observation *i* tend to vary systematically in size with the errors for other nearby observations *j* so that smaller/larger errors for *i* would tend to go together with smaller errors for *j*. Such clustering of residuals violates the assumption that the error terms are independent of one another.

What are the consequences of spatial correlation among the error terms, and what are the implications if we run an OLS assuming that observations are independent? If $\lambda \neq 0$, then the OLS coefficient estimates ignoring the spatial correlation would still be unbiased. However, the standard errors of the coefficient estimates would be wrong. Recall that OLS relies on an estimate of the variance that assumes independent observations. If this is not correct, then the OLS estimate of the variance $\hat{\sigma}$ will tend to underestimate the actual variance, in a manner analogous to the case of serially correlated errors over time. This occurs because the estimate of the variance disregards the correlation between the error terms for nearby observations. Moreover, the estimated coefficients are not necessarily efficient estimates or "close" to the true values of the impact of the features that we are interested in. We will return to estimation of the spatial error model later, but we first turn to its interpretation and relationship to the spatially lagged *y* model.

The spatial error model and the spatially lagged *y* model may seem superficially similar, as each suggests spatial dependence between observations. However, the two model specifications actually have very different substantive implications. The spatially lagged *y* is a simultaneous model with feedback between the observations: the value of y_i influences the value of y_j—which will in turn influence the value of y_k, which in turn influences the value of y_i. As we saw in Chapter 2, different values of the independent variable for one observation *i* will propagate through the connected observations, and the net impact will depend on the impact of these differences on other connected observations, via the spatially lagged *y* term. In contrast, in the spatial error model, dependence enters in the specification only through the error terms. The absence of the spatially lagged *y* term here implies that differences in the independent variables in *i* do not have effects on outcomes in observations connected to *i*. Thus, in a spatial error model specification, the observations are related only due to unmeasured factors that, for some unknown reason, are correlated across the distances among the observations.

3.2 Maximum Likelihood Estimation of the Spatial Errors Model

In the case of the spatially lagged *y* model, the coefficient ρ for the spatial lag is a parameter explicitly of interest on the right-hand side. In the spatial error model, λ is a coefficient indicating the correlation of the residuals rather than a right-hand side covariate of explicit interest. If we are only interested in estimating the $\hat{\beta}$ for *x*, and disregard λ altogether, OLS estimates will be unbiased and consistent for the spatial error model, unlike what was the case for the spatially lagged *y* model. However, the reported standard errors will be incorrect and the estimated coefficients are not necessarily efficient. These problems can be addressed by using generalized least squares estimation techniques similar to generalized least squares estimates often used in the presence of temporal correlation, where one first estimates the serial correlation and then seeks to transform the data and purge the serial correlation so that the normal regression assumptions are satisfied. This is typically done using an MLE based on the eigenvalues of the spatial connectivity matrix.

The log-likelihood function for the spatially lagged error model is

$$\ln \mathcal{L}(\beta,\sigma,\lambda) = \ln\left|\mathbf{I}-\lambda\mathbf{W}\right| - N/2\ln(2\pi) - N/2\ln\left(\sigma^2\right) \\ -\left(y-\lambda\mathbf{W}y-\mathbf{X}\beta+\lambda\mathbf{W}\mathbf{X}\beta\right)'\left(y-\lambda\mathbf{W}y-\mathbf{X}\beta+\lambda\mathbf{W}\mathbf{X}\beta\right)/2\sigma^2.$$

As in the case of the log likelihood of the spatially lagged *y* model, we run into complications over the log of the determinant $|\mathbf{I}-\lambda\mathbf{W}|$, which is an *n*th-order polynomial that is cumbersome to evaluate. However, we can again rely on Ord's (1975) result that this determinant can be written as a function of the product of the eigenvalues ω_i of the connectivity matrix **W**, $|\mathbf{I} - \lambda\mathbf{W}| = \prod(1 - \lambda\omega_i)$. Because the eigenvalues ω_i can be determined prior to optimization, this step can be separated from the likelihood evaluation for the other parameters (Anselin, 1988; Bivand, 2002). These estimators are implemented in common software options, including spdep in R.

3.3 Example: Democracy and Development

To show an actual example of the spatial error model in practice, we first revisit our example from Chapter 2 on democracy and wealth. We use the same data as in Chapter 2 and refer to this for all details on variable construction. Table 3.1 displays three sets of estimates for the democracy and income example. The results of the third column show the estimates for a model allowing for spatially correlated error, while the estimates of the first and second columns repeat the OLS and spatially lagged y model estimates from Chapters 1 and 2.

The code to estimate the spatial error code in R is straightforward:

```
# data and variables as employed in chapter 2.
sem.fit <- errorsarlm(democracy ~ log(gdp.2002/population),
     data=sldv, nb2listw(nblist), method="eigen", quiet=FALSE)
summary(sem.fit)
logLik(sem.fit)
```

As we see in Table 3.1, the coefficient estimate for the effect of logged GDP per capita is considerably larger in the spatial error model than the corresponding coefficient for the spatially lagged y model, albeit not as large as the coefficient estimate that we see in the nonspatial OLS model. The intuition here is that the OLS model is likely to overestimate the immediate impact of GDP per capita, as it does not take into account the spatial clustering in democracy and GDP per capita among neighboring countries. In that sense the estimate is also less precise, and we can think of the spatial lag as an omitted variable in the OLS model assuming independent observations. In contrast, the spatial error model corrects for the positive spatial correlation in GDP per capita and democracy, and this correction reduces the estimated coefficient for the impact of GDP. However, the spatial error estimates assume a model where the only spatial dependence between observations stems from the errors or excluded factors not in the systematic component

TABLE 3.1
Democracy and Logged GDP Per Capita

Variable	*OLS*			*SLDV*			*SEM*		
	$\hat{\beta}$	$SE(\hat{\beta})$	*t Value*	$\hat{\beta}$	$SE(\hat{\beta})$	*z Value*	$\hat{\beta}$	$SE(\hat{\beta})$	*z Value*
Intercept	–9.69	2.43	–3.99	–6.20	2.08	–2.98	–7.49	3.07	–2.44
ln GDP per capita	1.68	0.31	5.36	1.00	0.28	3.59	1.39	0.38	3.66
$\hat{\rho}$				0.56	0.08	7.43			
$\hat{\lambda}$							0.58	0.08	7.60
N		158			158			158	
df		1			2			2	
Log likelihood		–513.62			–491.10			–491.53	

NOTE: SEM, spatial error model; SLDV, spatially lagged dependent variable.

of the model. In contrast, in the spatially lagged y model, some of the net impact of an increase in GDP per capita of country i will be realized through the feedback effect that the immediate effect on i will exert on its neighbor j and its impact on i through the spatially lagged term, as the resulting impact of the democracy score will influence other countries and feed through the system until some equilibrium is reached. Hence, the estimated coefficient for GDP per capita in the spatially lagged y model will seem to be smaller than the spatially correlated error model, as it reflects the immediate impact rather than the long-run, net "equilibrium" effect implied by the model.

3.4 Spatially Lagged y Versus Spatial Errors

Since the two spatial parameters, ρ and λ, here are large and vastly greater than their standard errors, we can safely conclude that there is considerable spatial dependence in these data and that a standard OLS regression that assumes independent observations will be misleading. However, this leaves us with the question of what is the better model, the spatially lagged y or the spatially correlated error model? It is difficult to discriminate between the spatially lagged y and a spatial error model purely on statistical grounds. The two models are not nested, so it is not possible to see one as a subset of the other, as is often the case with hypothesis testing where we impose additional restrictions on a model. It is possible to use formal tests for comparing nonnested models.[1] However, these will often be inconclusive and unlikely to suggest strong support for one model over the other. In this case, we can see that the log likelihoods for the two models are very similar, as the log likelihood for the spatially lagged y model is only marginally smaller than that of the spatially correlated error model. Since both models entail the same number of parameters, there is no clear basis for saying that one is more parsimonious than the other and, hence, there is little empirical guidance to tell the two models apart in terms of fit to the data alone. One approach to follow would be cross-validation or out-of-sample prediction tests, but these approaches are beyond the scope of this book.

More important, whether the spatially lagged y model or the spatial error model is most appropriate is really a prior theoretical question, which must be considered relative to the goals of a specific research question. If we expect to see, or are interested in, feedback, then the spatially lagged y model would seem a more appropriate model. In our democracy example, it seems quite reasonable to expect that a country's level of democracy is affected by the extent to which other countries are democratic (see, e.g., Gleditsch, 2002a; Gleditsch & Ward, 2007). In contrast, it seems much less plausible to assert that there is no diffusion effect from levels in other states per se but rather some other omitted feature from the systematic component that induces spatial correlation in the errors of the model. Hence, in this case we believe that the spatially lagged y model is more appropriate than the spatial error model.

More generally, the spatial error model is probably less interesting for applications in the social sciences, and in our view, the spatial error model is appropriate primarily when researchers believe that there is some spatial pattern that will be reflected in the error term, but they are either unwilling or unable to make assumptions about the origin of the error. The reason for this is that most models proposed in the social sciences do not do a great job of specifying attributes of the individual observations that fully capture the observed spatial clustering. As a result, there is still a lot of leverage in undertaking that specification in the context of a spatially lagged dependent variable. If, however, we had an area in which most of the important mechanisms were understood and fully specified in the systematic component of the model and if there were still some correlation in the error terms, it would be useful to employ a spatial error model to correct for this residual nuisance. On the whole, since social science modeling is typically characterized by little attention to dependencies among the data, the spatial error can often provide a substantial improvement.

3.5 Assessing Spatial Error in Dyadic Trade Flows

To illustrate a case where the spatial error model seems more appropriate, we consider an application to the study of dyadic trade flow. A dyad is a pair of two units, where the response may be some measure of trait or interaction between units, in our case the volume of trade between two states i and j. In some cases, we may wish to distinguish the direction of interaction between i and j, which could be denoted $i \rightarrow j$ for behavior of i toward j. Alternatively, undirected interactions might be denoted with a subscript $i \leftrightarrow j$. A system of n units will give rise to $n \times (n-1)$ directed dyads, or $[n \times (n-1)]/2$ undirected dyads if we do not distinguish the direction of flows or interaction between the units. Outside our example of trade here, dyadic observations are very common in international relations, where we may be interested in estimating how some feature affects the likelihood of a particular event or behavior, such as conflict between two states i and j.

The conventional approach in dyadic analysis in international relations is to treat interaction as a function of characteristics of dyads or the two units they are composed of and to take the individual dyads as independent of one another once we have taken into account the relevant explanatory factors. The spatial error model can be useful in addressing possible dependence between such dyadic observations.[2]

Gold-standard models of international trade have not changed substantially since they were first introduced: They are based on an analogy to Newtonian models of gravity. Trade is a function of the economic mass of the trading countries but is inversely proportional to their "distance." Existing empirical work has suggested a number of factors likely to influence the extent of trade between states i and j. The workhorse of empirical trade model is the so-called gravity model of trade, which postulates that the volume of trade between two countries ($T_{i \rightarrow j}$) should be proportional to the product of their mass, in terms of the size of their economies (GDP_i and GDP_j) and population (P_i and P_j), and the geographical distance separating the two states ($D_{i \leftrightarrow j}$). The model is usually stated as an additive model in logarithmic form:

$$\log(T_{i \rightarrow j}) = \alpha + \beta_1 \ln(GDP_i) + \beta_2 \ln(GDP_j) + \beta_3 \ln(P_i) + \beta_4 \ln(P_j) + \beta_5 \ln(D_{i \leftrightarrow j}) + \epsilon,$$

where the "mass" coefficients ($\beta_1, \ldots, \beta_4$) are expected to be positive and the distance coefficient (β_5) negative. Feenstra, Rose, and Markusen (2001) and Rose (2004) provide examples of recent applications.

This core gravity model has no political content, but many social scientists have been interested in how political factors may influence trade flows. For example, Pollins (1989a, 1989b) argues that political relations are likely to exert a strong influence on trade volumes, as countries will be less likely to have high volumes of trade with countries with which they have otherwise poor political relations, either because traders fear political disruption or because governments will impose restrictions on trade with antagonistic states. Morrow, Siverson, and Tabares (1998) hold that democracies are likely to trade more with other democracies and that military conflict should be associated with less trade than would otherwise be expected. Both empirical analyses suggest that these features influence trade flows.

One issue that has received little attention in research on trade is the potential problems that the dyadic observations may not be independent of one another. Although there has been a great deal of research on the potential problems that subsequent observations for the same dyad over time may not be independent (see, e.g., Beck & Katz, 1996), most studies have assumed that observations of different dyads at the same point in time can be considered independent of one another. However, there are many reasons to expect that this may not hold in the case of trade flows. Dyadic data will tend to have a complex structure, since the same state will enter into a very large number of dyads. First, the trade flows $T_{i \to j}$ and $T_{i \to k}$ cannot be considered independent of one another because they share the same sender. Second, it will often be the case that the flow from one state i to j $(T_{i \to j})$ will be positively correlated with the reverse flow from j to i $(T_{i \leftarrow j})$. Higher-order dependencies are also often found in such data.[3] Economists typically average the values of $T_{i \to j}$ and $T_{j \to i}$ and analyze the decomposition of this triangular matrix. Such procedures guarantee even more dependence among observations. Moreover, it is widely known that most reported trade data rely on estimates that appear to be based on imputations from other dyadic flows (e.g., Rozanski & Yeats, 1994). Such imputations can induce serial correlation in the figures. For example, the trade data reported by the World Bank show significant divergences from the distribution one would expect from Benford's law of the distribution of first digits, which is a common test for the quality of data and possible evidence of made up figures.[4]

This is a case where the spatially lagged error model is appropriate, as we would expect the error terms of particular dyads to be linked, rather than their observed trade flows. The net volume will depend on the mass of the countries in a dyad, but conditioning on this alone will not take into account the variation of the errors due to dyadic dependencies. We have previously talked about distance and connectivity in terms of the geographical distance between two units. In this case, we have suggested a dependence structure deriving from dyads sharing a common member, where the structure of dependence is not "spatial" in the conventional sense. However, there is nothing that precludes spatial ideas from being applied to nongeographical concepts of distance. In this case, we can devise a weighting scheme where other dyads are considered connected to a particular dyad $i \to j$

if they contain at least either i or j. See Beck, Gleditsch, and Beardsley (2006), Deutsch and Isard (1961), and Lofdahl (2002) for further discussion on alternative concepts of "distance."

For an empirical application, we consider data on trade among European and African dyads from Gleditsch (2002b). Specifically, we look at the exports from country i to country j, denoted $T_{i \to j}$. The samples for Africa and Europe provide an interesting comparison with likely variation in the quality of the data, as we would expect the accuracy of the European trade data to be much higher than the data available for African states, given the differences in infrastructure and capacity for monitoring economic activities. All the trade data in our examples are from 1998. In our sample, "observed" data in the sense of trade flows reported by the International Monetary Fund and other international agencies comprise about 75% of all the European dyads (i.e., data with origin codes 0 or 2 in the Gleditsch 2002b data). For Africa, however, relying on officially reported figures would leave us with trade flow data for only 15% of the dyads. In our examples, we will use only the officially reported data for Europe, but employ all estimates from all sources, including the potentially more contentious estimates, for the analysis of trade flows in Africa.

The standard gravity model variables include the size of the economies and the population of the two member states (data from Gleditsch, 2002b) and the distance between their capital cities. In addition, our model draws on the existing literature on the political determinants of trade and include the similarity of political orientation by the S similarity score of two countries' UN voting records (see Gartzke, 1998; Signorino & Ritter, 1999). Our measure of democracy is taken from the POLITY IV data. We use a modified version, including estimates for countries not included in the original POLITY data based on the Freedom House data.[5] We use the lower of the two values on the 21-point institutionalized democracy scale suggested by Jaggers and Gurr (1995), rescaled so that all the values are positive. Finally, we consider whether the two states in a dyad are involved in a militarized interstate dispute (see Jones, Bremer, & Singer, 1996).

```
source("chapter3data.R")
tab3.sem <- errorsarlm(logtrade ~ logdem + logapop + logbpop +
      logargdppc + logbrgdppc + logs + logdist + logmid,
      data=logdat98,na.action=na.omit,
      nb2listw(dlist,style="W"), method="eigen")
summary(tab3.sem)
logLik(tab3.sem)
```

Tables 3.2 and 3.3 display OLS and spatially correlated error model estimates for trade flows among European and African dyads, respectively. As suggested by $\hat{\lambda}$, there is strong evidence of positive spatial correlation among dyads in both the African and the European samples. Moreover, by comparing the OLS and the spatial error model estimates we can see that the point estimates for the magnitude of some effects highlighted in the literature concerned with the political determinants of trade change notably when we take into account spatial correlation among the residuals based on the common membership structure of the dyads rather than treating them as independent observation. In particular, the coefficient estimate for the negative effect of an MID declines by almost 25% in the European sample and by over 40% in the African sample when we take into account the spatial correlation among dyads. The coefficient estimate for democracy is reduced to about a quarter of its original size in the African sample, while in the European sample it increases by almost 15%. Moreover, the standard errors for the individual coefficient estimates are generally larger for the spatial error model than for the OLS model, indicating that a model that treats the individual dyadic observations as independent of one another risks inducing overconfidence in the estimates through incorrect standard errors. More generally, although we do not have any coefficients that switch from being "significant" to "insignificant" based on conventional thresholds, it can certainly

TABLE 3.2
Exports, Europe: $T_{i \to j}$

Variable	*OLS*			*SEM*		
	$\hat{\beta}$	*SE*($\hat{\beta}$)	*t Value*	$\hat{\beta}$	*SE*($\hat{\beta}$)	*z Value*
Intercept	−32.70	0.67	−48.82	−33.94	1.71	−19.90
log Democracy	0.38	0.06	5.93	0.43	0.10	4.38
log Population *i*	0.86	0.02	40.37	0.89	0.03	31.46
log Population *j*	0.75	0.02	34.93	0.77	0.03	27.33
log GDP per capita *i*	1.54	0.04	35.23	1.56	0.06	17.35
log GDP per capita *j*	1.01	0.04	23.07	1.03	0.06	7.66
log *S*	0.33	0.05	6.92	0.35	0.05	7.69
log Distance $i \leftrightarrow j$	−0.34	0.01	−24.33	−0.34	0.01	−25.83
log Dispute $i \leftrightarrow j$	−1.94	0.27	−7.14	−1.48	0.29	−5.01
$\hat{\lambda}$				0.98	0.01	73.73
N		1,500			1,500	
df		8			9	
Log likelihood		−2324.8			−2239.668	

NOTE: SEM, spatial error model.

TABLE 3.3
Exports, Africa: $T_{i \to j}$

Variable	OLS			SEM		
	$\hat{\beta}$	$SE(\hat{\beta})$	*t Value*	$\hat{\beta}$	$SE(\hat{\beta})$	*z Value*
Intercept	–7.41	0.33	–22.38	–7.47	1.45	–5.16
log Democracy	–0.04	0.04	–1.08	–0.01	0.05	–0.15
log Population i	0.26	0.01	20.51	0.26	0.02	14.45
log Population j	0.23	0.01	17.81	0.23	0.02	12.55
log GDP per capita i	0.38	0.02	17.96	0.38	0.03	12.78
log GDP per capita j	0.31	0.02	14.82	0.31	0.03	10.55
log S	3.41	0.40	8.50	3.43	0.47	7.24
log Distance $i \leftrightarrow j$	–0.17	0.01	–20.81	–0.17	0.01	–22.21
log Dispute $i \leftrightarrow j$	–0.71	0.18	–3.85	–0.42	0.18	–2.37
$\hat{\lambda}$				0.99	0.01	124.2
N		2,550			2,550	
df		8			9	
Log likelihood		–3096.2			–2945.9	

NOTE: SEM, spatial error model.

be the case that many apparent findings from analyses treating individual dyads as independent of one another may turn out to be less robust when we take into account spatial dependence among the observations.

3.6 Summary

In this chapter, we have introduced the spatially correlated error model of spatial dependence. Since it will generally be difficult—if at all possible—to determine whether the spatially correlated error or the spatially lagged y model will be appropriate based on statistical criteria alone, researchers should think carefully about which of the two would provide the most plausible way to incorporate spatial dependence. We have argued that although the spatially lagged y model is appropriate when we expect nearby values of the response to exert a direct effect on the value of the dependent variable for a unit, the spatially correlated error model is appropriate when we believe that some unobserved feature not included in the systematic part of regression model may lead to a spatially correlated pattern in the errors of the model. Our example of dependence among dyads, where a single state enters into a large number of different dyadic observations, also illustrates how the concept of spatial dependence can be extended to distance concepts based on metrics other than geographic distances.

Notes

1. See, for example, Clarke (2001) for a discussion of nonnested tests.

2. A more complete examination of these and other dependencies can be found in Ward and Hoff (2007); for binary dependent variables, see Ward, Siverson, and Cao (2007).

3. Wasserman and Faust (1994) provide an overview of the triadic implications in such dyadic data, a staple of social network analysis.

4. The first-digit law, named after physicist Frank Benford, states that the leading digit in data will follow a law where 1 will be the most frequent leading digit and larger numbers will become successively less common, or more precisely, the frequency of a digit p is closely approximated by $\log(p+1) - \log p$. This law applies to a large distribution of naturally occurring data, and it has been suggested that large deviations from this distribution could be used as evidence of poor-quality data or fraud. We refer to Varian (1972) for a discussion of Benford's law.

5. See http://privatewww.essex.ac.uk/~ksg/polity-data.html.

4. EXTENSIONS

The previous chapters in this book suggested the necessity and benefits of taking into account spatial patterns in the analysis of social science data. We illustrated important ways in which this can be done within a familiar linear regression framework—namely, the *spatially lagged dependent variable model*, where the value of y_i in connected units exerts an impact on y_j, and the *spatial error model*, in which there is simply a spatial correlation of the errors for connected observations. These two are widely used spatial regression methods and are useful for many applications. However, there are many additional varieties of spatial regression models and extensions to other settings that we have not considered, and our attention has been limited to cross-sectional data for continuous dependent variables. In this section, we outline some extensions of spatial statistical models and some of the thorny issues that spatial analysts may face. Although our overview must be brief and we do not cover actual examples of these alternative extensions and alternative approaches, we provide suggestions for further readings. Bivand, Pebesma, and Gomez-Rubio (forthcoming) provide additional pedagogical materials sculpted for the R statistical package.

4.1 Specifying Connectivities

One key problem facing analysts is how to construct and treat connectivities among observations. Most applications of spatial regression models presume prespecified delineations of connectivities between observations. They should be based on theory or hunches about the likely ways in which observations are substantively dependent. In practice, this is often based on convenience or common approaches thought to be state of the art. Researchers should be aware that the ways in which connectivities are chosen and coded can give rise to different views of the world. It is trivial that different results may create variation in the direct specification of connections between units. However, more subtly, these choices also affect the spatial structure implied by the spatial multiplier as well as the modeled covariance structure (Wall, 2004). Even in the case of connectivities said to be based on geographical distance, the same spatial topology can generate different connectivity structures depending on the researcher's decisions. To see this, consider the differences between three common spatial encodings: *rook* (common boundaries), *bishop* (common vertices), or *queen* (both boundaries and vertices) in a partial map of the United States shown in Figure 4.1. Colorado and Utah are neighbors by sharing a common border, by sharing

Figure 4.1 The Four Corners Area of the United States.

vertices, and by sharing both a border and vertices. Colorado and Arizona, on the other hand, do not share borders, but do share a "vertex." There is only one example of which we are aware of a similar situation in the political map of country boundaries in the contemporary world: the Caprivi Strip in South West Africa.

More typically, scholars identify units as *close* if they are within some distance threshold, based on minimum distances among administrative centers or geographical centroid or midpoints. Gleditsch and Ward (2001) discuss some of the problems of common midpoint measures, which may be far from the boundaries of large units in the case of administrative centers and outside the boundaries of a unit for strangely shaped units. An overly narrow threshold may create a large number of islands, an issue we raised in Chapter 1 in the context of New Zealand. It is about 4,100 km from Alice Springs, Australia, to Christchurch, New Zealand, approximately the same as the distance from Paris to Dar es Salaam. This implies that choosing a general distance band to link the centroids of Australia and New Zealand and then applying the same criterion to other pairs of countries would make most African countries and many countries in the

Middle East and Asia direct neighbors of France. Overly broad demarcations result in connecting almost everything to everything else. Figure 4.2 illustrates the dramatic increase in density in a graph linking states if their centroids are within 4,000 km (Panel b) compared with the graph linking centroids only if they are located within 400 km of one another (Panel a). Ad hoc decisions to add specific connectivities or to use algorithms that choose the k nearest neighbors of each observation raise questions of why the same criteria are not applied to other cases. However, such ad hoc decisions may be useful, even necessary, in applied research. In summary,

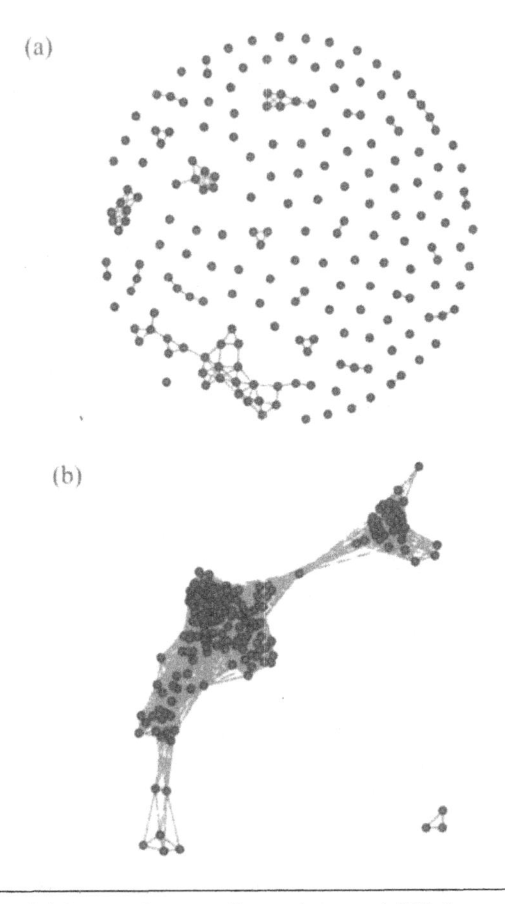

Figure 4.2 Linkages Among Countries, at 4,000 km and 400 km, Using Centroid Distances: (a) 400 km; (b) 4000 km.

the choice of connectivity encoding always has substantive implications for empirical results because diffusion through differing networks leads to different conclusions.

Connectivities based on nongeographical measures, such as trade flows, may create additional problems. More specifically, if the nongeographic distance measures are based on variables that are also included in the actual spatial model, then the connectivities may not be exogenous, resulting in identification, as well as estimation, challenges. Researchers should devise connectivity matrices that plausibly match the spatial interaction process being studied. Although goodness of fit and cross-validation can be helpful in eliminating poor choices, devising connectivities is a theoretical issue and there are no simple diagnostics or heuristics that uniquely define the "right" approach. We also stress that the problems in identifying connectivities create hurdles for testing a null hypothesis of the absence of spatial dependence, since the null can only be rejected relative to a specific set of connectivities.

4.1.1 Handling Connectivities

Another issue pertains to how, once specified, the connectivities should be handled in the analysis itself. Should all connectivities be given equal weight, or should we weigh some observations differently, for example, by some measure of size or importance? In the examples we have examined in this book, we have assumed that Russia and Estonia carry equal weight for countries connected to both. However, there is nothing that requires equal weighting of all connectivities. Researchers may wish to experiment with alternative weighting schemes if this makes sense in the context of their specific research questions.

We have only considered regression models with a row-normalized matrix, **W**, specifically where the sum of all connectivity weights add to 1. This specific normalization has the advantage that the spatial lag y^s will have the same potential metric or units as y itself. However, whether normalization makes sense in specific applications depends on the problem at hand. Murdoch et al. (1997), for example, are interested in how a country's emissions of pollutants are influenced by depositions from other countries. The relevant issue concerns the total amount of emitted pollutants; normalizing the connectivity matrix by the number of connected countries is probably not appropriate.

Analysts should treat conventions in the spatial statistical literature as suggestions and carefully consider whether they are sensible in their specific research setting. It is generally useful to explore several plausible alternatives.

4.1.2 One Versus Many Connectivities

So far, we have explored cases with a single spatial dependence term, represented in a single connectivity matrix. In many cases, there may be several possible networks or forms of dependence. Often, it makes sense to include alternative delineations of connectivity based on geographical distance or other political networks, such as trade interaction, cultural similarity, or the ethnicity or profession of individuals (see, e.g., Beck et al., 2006; Lacombe, 2004; Lin et al., 2006). Direct impacts may arise not only

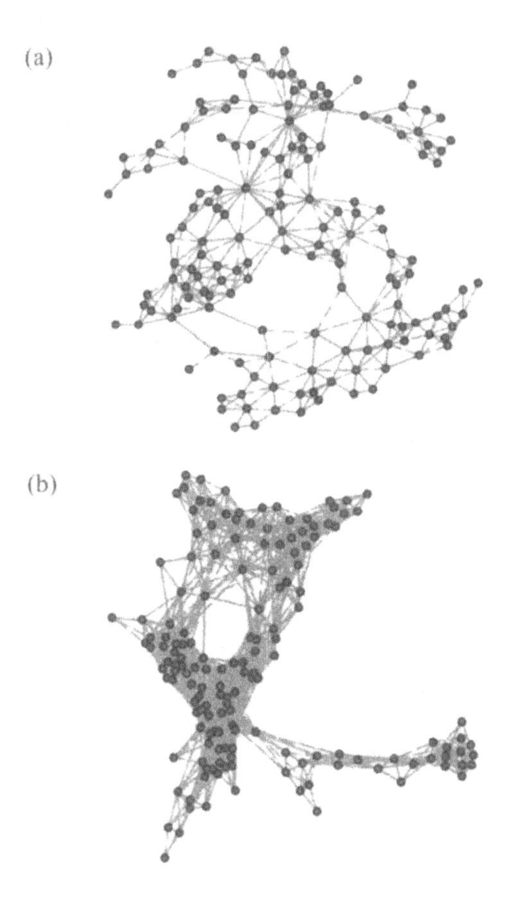

Figure 4.3 First- and Second-Order Connectivities for 158 Countries, Based on Nearest Neighbor Distances of 200 km: (a) First Order; (b) First and Second Orders.

from first-order connectivities but also from higher-order connectivities. Figure 4.3 illustrates the first- and second-order connectivities from the 158-country example used in the previous chapters.

It is possible to generalize the spatially lagged y model to include two (or more) distinct connectivity matrices, $\mathbf{W}^A$ and $\mathbf{W}^B$, and estimate separate parameters ρ_1 and ρ_2 for the relative impact of each by

$$y_i = \mathbf{x}_i\beta + \rho_1\mathbf{w}_i^A y + \rho_2\mathbf{w}_i^B y + \epsilon.$$

The expanded spatially lagged y model is more complicated to estimate than the standard spatial autoregressive model. Provided the two matrices are sufficiently different and do not contain entirely overlapping information, this model can be estimated. If the matrices are too similar, problems arise that resemble those caused by collinearity in the classical regression model. The MLE discussed previously can be generalized to this case (although this is not yet implemented in R). This model can also be estimated by instrumental variables.

4.2 Inference and Model Evaluation

As is commonly the case with much data in the social sciences, spatial data do not come from a random sample. Spatial analysis requires a relatively complete spatial coverage because analysis of data with many missing units can lead to nonsensical inferences about spatial clustering or the impact of nearby units. Classical inference is based in large part on asymptotic assumptions that may be difficult to justify in most spatial contexts. These require, in essence, that the number of neighbors does not cascade upward as a function of the size of the area that is examined. Even so, the data typically involved do not resemble a sample but tend to be a cross section of some area or universe of interest. This results in an empirical search for models that seem plausible for the process under investigation and that, in principle, could have generated the observed data. Classical approaches may be based on the notion of a generalizing to the so-called superpopulation, of which the observed spatial pattern is realization, but this concept does not sit so well with spatial analysis where one typically studies what Berk, Western, and Weiss (1995) call "apparent populations" (see also Leamer, 1978).

One possible solution to this conundrum is to treat estimation more heuristically and use cross-validation of the results to examine the performance of the estimated model on data that have not been used in the spatial regression estimation following the tradition of Geisser (1974, 1975). In the spatial context, this could be accomplished by using observed data from a subsequent time period or a different spatial domain. Bivand (2002), for example, splits data into two geographical areas to evaluate the performance of different modeling approaches in terms of their ability to predict observations in the other half of the data.

4.2.1 Discrete and Latent Variables

We have assumed that y can be treated as a continuous variable. Yet many phenomena of interest to social scientists are discrete events, which are observed either in binary form or as counts. Alternatively, these might be outcomes of partially observed latent processes. Just as linear regression is suboptimal for these data, the models examined here are not generally appropriate for such data. However, it is possible to generalize the idea of a lagged dependent variable or autoregressive process to binary event or count data, including, for example, the autologistic model in which values of y for nearby units influence $Pr(y_i = 1)$ (Besag, 1972, 1974; Christensen & Waagepetersen, 2002; Huffer & Wu, 1998; Ward & Gleditsch, 2002). Estimating these models is more difficult than the continuous case due to an intractable likelihood resulting from the fact that y appears on both sides of the equation. Traditional approaches have treated the y in connected observations as fixed for estimation purposes, but modern computing power allows approximating the full likelihood through simulation (Geyer & Thompson, 1992).

4.2.2 Spatial Heterogeneity

Normally, the major effects investigated with regression are *fixed effects* and pertain to the relationships between the independent and dependent variables *everywhere*. However, if this relationship is different in one part of the world than it is in another, we have a form of *spatial heterogeneity* where the effects are geographically conditional. Spatial heterogeneity is an opportunity to learn something about the phenomenon of interest as well as a curse. On one hand, it offers another way to disaggregate regression results so that they are sculpted to be more revealing in differing regions. On the other hand, it causes havoc with standard regression assumptions of constant variance across the domain of analysis. Geographically weighted regression, known as GWR, is a windowing technique for exploratory data analysis that provides estimates of regression coefficients *for each geographical location*, based on a weighting of other observations near that location. This approach to spatial analysis is developed in Brundson, Fotheringham, and Charlton (1996) and more fully detailed in Fotheringham, Charlton, and Brundson (2002). A recent example in political science is Calvo and Escolar (2003); an interesting application in demography is found in Işik and Pinarcioğlu (2007).

4.2.3 Point and Geostatistical Data

The approaches explored so far treat geography as being compartmentalized. Countries, for example, are taken as a lattice, implying that each country has a location somewhere on a grid but that no country takes up more than one grid location. This approach is useful for many kinds of data but not all phenomena are reasonable to observe for area or lattices and data are often not organized in this fashion. Indeed, many types of data are organized as point data in a geo-referenced manner so that the exact or approximate location of each observation is observed in a continuous topology, not as an imaginary grid. Geostatistical methods attempt to model the spatial covariation to construct a geostatistical surface for the continuous geography based on the information that exists at specific locations within that geography. One approach to doing this is known as Kriging, formally developed by Matheron (1963) but named after the South African mine engineer Danie G. Krige, who pioneered plotting distance-weighted average gold grades.[1] This approach is widely employed in the geophysical sciences and has recent applications in the social sciences as well (Cho & Gimpel, 2007). Although data traditionally only have been available for large aggregate units or without spatial identifiers, there is an increasing availability of geographically disaggregated data or explicitly geo-referenced data.

4.2.4 Hierarchical Models

Following the early contribution of Besag (1974), there has also been considerable work on models that are conditionally autoregressive, often called CAR models. In a conditional model, the random variable observed at a certain location is conditioned on observations at neighboring observations, which are treated as exogenous. In multivariate and hierarchical models, not only are spatial lags taken as exogenous but so are other explanatory variables. There is considerable work underway to exploit this approach, sometimes focusing on several responses or dependent variables. Recent work on these topics can be found in Jin, Banerjee, and Carlin (2007) and Rue and Held (2005).

A related approach to modeling spatial variation is to examine the sources of local variation in a hierarchical fashion. Hierarchical spatial models are based on incorporating different sources of uncertainty from different levels of analysis. These models result in linking together several levels of analysis through the use of probability distributions. In our example of development and democracy, these levels might include the following: (1) local within-country variation in the characteristics of factions and institutions that help determine the daily ebb and flow of politics and economics; (2) neighborhood effects from very nearby countries that have strong connections to and impacts on particular countries; (3) regional sources of variation that operate on a broad set of countries, including organizations organized along regional lines; and (4) global forces that affect every country to some extent, as seen in global markets for certain commodities. Models that explicitly explicate each of these sources of variation will be hierarchical models.

Recent work to develop this perspective is based on a Bayesian approach that is itself dependent on the use of iterated approaches (Markov chain Monte Carlo, Gibbs sampling, Metropolis-Hastings, etc.) to implement a strategy of obtaining distributions of parameters at all the levels of the spatial process. Such models require intense computation but are very promising. The recent R package `spBayes` can also help in facilitating Markov chain Monte Carlo computations for univariate and multivariate spatial models (Finley, Banerjee, & Carlin, 2007). Waller, Carlin, Xia, and Gelfand (1997) is an influential application, and Banerjee et al. (2004) provide a good overview of the hierarchical approach.

4.2.5 Time Series Data

We have discussed estimation of models for a cross section of observations at the same time period. Much of social science analysis is based on time series cross section (TSCS) data structures, where the same unit is observed at several different time periods. The spatially lagged y model can be generalized for TSCS data as

$$y_{i,t} = \mathbf{x}_{i,t}\beta + \rho \mathbf{w}_i y_{i,t} + \epsilon_{i,t}.$$

This model is likely to suffer from problems of serial correlation over time, since $y_{i,t}$ is likely to be very similar to $y_{i,t-1}$, which in turn creates problems with assumption of independent errors. One way to address this is by adding a temporal lag of y to the model, yielding

$$y_{i,t} = \phi y_{i,t-1} + \mathbf{x}_{i,t}\beta + \rho \mathbf{w}_i y_{i,t} + \epsilon_{i,t}.$$

It is difficult to estimate TSCS models with simultaneous spatial dependence if we need to simultaneously account for both temporal and spatial dependencies. If we add the lagged dependent variable to the right-hand side, the Jacobian of the transformation of the error ϵ to y becomes considerably more complicated, and as far as we know, no one has come up with a satisfactory estimator for this model. However, if one is willing to assume that the influence of $y_{i,t}$ on the neighboring y's occurs with a one time period lag (i.e., $y_{i,t-1}$), it is possible to use OLS because the relevant neighboring values of y can be treated as predetermined at time t. This is simply

$$y_{i,t} = \phi y_{i,t-1} + \mathbf{x}_{i,t}\beta + \rho \mathbf{w}_i y_{i,t-1} + \epsilon_{i,t}.$$

Assuming that the spatial effects enter with a time lag will often be as plausible as assuming an instantaneous effect. Moreover, it is possible to test to what extent a model succeeds in accounting for both spatial and temporal dependencies by conducting the appropriate tests with the estimated residuals from the model and especially by employing cross-validation and out-of-sample heuristics (for further discussion, see Beck et al., 2006).

4.3 Summary

Spatial dependence plays a big role in many social phenomena. Taking spatial aspects into account in our analyses is entirely feasible, but does require some additional assumptions and information. Because statistical and computational developments have reduced the barriers to undertaking spatial analysis of social data, we should expect new insights about the social and spatial processes that interest social scientists. Our own experience has convinced us that social science data are characterized by many unexplored dependencies. Taking even some of these into account generally yields important, new insights.

Note

1. Following the name of its originator, kriging should be pronounced "kricking."

APPENDIX: SOFTWARE OPTIONS

For a long time, spatial estimators were not available in standard statistical packages, forcing interested researchers to either write their own program or purchase Anselin's *SpaceStat* software. This situation has changed markedly over the last couple of years. In this section, we review the available options.

Many of the software options still rely on the Ord approach and evaluate the eigenvalues of **W** prior to optimization, although some alternatives now use the faster Pace and Barry approach. Many software packages require a full $n \times n$ matrix as input. This tends not to work well for large data sets. Since there usually tend to be many 0 entries in connectivity matrices, software options that can use sparse matrix representations allow for work on much larger data sets.

We list here some software options for spatial analysis.

1. Anselin's *SpaceStat* software is no longer maintained as a separate package under the control of Anselin but has been purchased by a commercial company and is available as a part of a geographic visualization program *TerraSeer* (see http://www.terraseer.com/products_spacestat.php). The cost of the software is substantial, even for an Academic license. The old version of *SpaceStat* runs in MS-DOS and has what might be described as an old-school, menu-driven interface. It also relies on the Ord approach to determinants of the weights matrix and requires a full matrix representation in estimation. We are not familiar with the current product distributed as part of *TerraSeer* and thus cannot comment on how this may differ from previous versions.

2. Anselin and colleagues have developed a new package called *GeoDa*, available at https://www.geoda.uiuc.edu/. GeoDa does both exploratory spatial data analysis and simple spatial regression analysis. GeoDa is completely driven by a point-and-click interface and does not require any programming; however, it does not allow users to customize or modify any of the features as would be possible in a general statistical package. Anselin, Syabri, and Kho (2004) suggest that it is primarily appropriate as a learning package, "with more sophisticated users 'graduating' to R after being introduced to the techniques in GeoDa" (p. 3).

3. Pisati's `spatreg` macro for *Stata* allows estimating the spatial autogregressive and error models. The downside of this program or macro is that it relies on the Ord approach and requires a full matrix specification. The standard Intercooled version of *Stata* also has a restrictive maximum size for matrices. See Stata technical Bulletin `sg162`. See `help stb` within *Stata* for help on installation.

4. Roger Bivand has developed an R-package (*spdep*) that implements the models discussed in this book. This package also allows for a sparse list representation of connectivity matrices. Bivand has also developed various material for integration of R and GRASS, an open source GIS program. Moreover, many utilities have been made available for creating maps and extracting information from *Arcview*'s shapefiles format in R. See http://cran.r-project.org/src/contrib/Descriptions/spdep.html for further details on the package. These, as well as the underlying software platform, are open source and available free of charge.

5. Several *MATLAB* toolboxes for spatial analysis are available. Pace and Barry's `Spatial Statistics` is available free of charge at http://www.spatial-statistics.com/. MATLAB itself is not free of charge. LeSage's `Spatial Econometrics` toolbox, available at http://www.spatial-econometrics.com, is particularly helpful in estimating models on very large data sets and also allows estimating a spatial autoregressive model with two connectivity matrices through the `saw()` command.

6. The latest version of the commercial package *ARCINFO* from ESRI, Inc., includes many facilities for undertaking statistical analysis of spatially organized databases, especially those in its `Statistical Analyst` toolkit. It is especially adept at calculating neighborhood statistics and classification analysis.

7. Splus is available from Insightful Corp. Like R, it is also based on the S statistical language. It includes a module (SpatialStats) that features many analysis tools for spatially correlated data. It has tools for geostatistical, point, and lattice spatial data.

8. WINBUGS (http://www.mrc-bsu.cam.ac.uk/bugs/) and GeoBUGS are two programs oriented to Bayesian analysis. GeoBUGS was developed by an epidemiologist as an add-on to WinBUGS. It supports the Bayesian analysis of (relatively small) spatial models.

9. Schabenberger and Gotway (2005) provide an extensive set of macros and programs to undertake spatial data analysis in SAS, available at the publisher's Web site: www.crcpress.com

10. Spatial hierarchical methods are readily available through the R package `spBayes`, facilitating the MCMC computations often required in such models (Finley et al., 2007).

REFERENCES

Adolph, C. A. (2004). *The dilemma of discretion: Career ambitions and the politics of central banking*. Unpublished doctoral dissertation, Harvard University, Cambridge, MA.

Anselin, L. (1988). *Spatial econometrics: Methods and models*. Dordrecht, The Netherlands: Kluwer.

Anselin, L. (1995). Local indicators of spatial association-LISA. *Geographical Analysis*, *27*, 93–115.

Anselin, L., Syabri, I., & Kho, Y. (2004). *GeoDa: An introduction to spatial data analysis (Typescript)*. Urbana-Champaign: Department of Agricultural and Consumer Economics, University of Illinois.

Banerjee, S., Carlin, B. P., & Gelfand, A. E. (2004). *Hierarchical modeling and analysis for spatial data*. Boca Raton, FL: Chapman & Hall.

Baybeck, B., & Huckfeldt, R. (2002). Urban contexts, spatially dispersed networks, and the diffusion of political information. *Political Geography*, *21*, 195–220.

Beck, N., Gleditsch, K. S., & Beardsley, K. (2006). Space is more than geography: Using spatial econometrics in the study of political economy. *International Studies Quarterly*, *50*, 27–44.

Beck, N., & Katz, J. N. (1996). Nuisance vs. substance: Specifying and estimating time-series—Cross-section models. *Political Analysis*, *6*, 1–36.

Berk, R. A., Western, B., & Weiss, R. E. (1995). Statistical inference for apparent populations (with discussion). *Sociological Methodology*, *25*, 421–485.

Besag, J. E. (1972). Nearest-neighbour systems and the auto-logistic model for binary data. *Journal of the Royal Statistical Society, Series B, Methodological*, *34*, 75–83.

Besag, J. E. (1974). Spatial interaction and the statistical analysis of lattice systems (with discussion). *Journal of the Royal Statistical Society, Series B, Methodological*, *36*, 192–225.

Bivand, R. (2002). Spatial econometrics functions in R: Classes and methods. *Journal of Geographical Systems*, *4*, 405–421.

Bivand, R., Pebesma, E., & Gomez-Rubio, V. (forthcoming). *Applied spatial data analysis with R*. New York: Springer.

Boots, B. N., & Tiefelsdorf, M. (2000). Global and local spatial autocorrelation in bounded regular tessellations. *Journal of Geographical Systems*, *2*, 319–348.

Brundson, C., Fotheringham, A. S., & Charlton, M. (1996). Geographically weighted regression: A method for exploring spatial nonstationarity. *Geographical Analysis*, *28*, 281–298.

Burkhart, R., & Lewis-Beck, M. (1994). Comparative democracy: The economic development thesis. *American Political Science Review*, *88*, 903–910.

Calvo, E., & Escolar, M. (2003). The local voter: A geographically weighted approach to ecological inference. *American Journal of Political Science*, *47*, 189–204.

Cho, W. K. T., & Gimpel, J. G. (2007). Prospecting for (campaign) gold. *American Journal of Political Science*, *51*, 255–268.

Christensen, O. F., & Waagepetersen, R. (2002). Bayesian prediction of spatial count data using generalized linear mixed models. *Biometrics*, *58*, 280–286.

Clarke, K. A. (2001). Testing nonnested models of international relations: Reevaluating realism. *American Journal of Political Science*, *45*, 724–744.

Cleveland, W. S. (1993). *Visualizing data*. Summit, NJ: Hobart Press.

Cliff, A. D., & Ord, J. K. (1971). Evaluating the percentage points of a spatial autocorrelation coefficient. *Geographical Analysis*, *4*, 51–62.

Cressie, N. A. C. (1993). *Statistics for spatial data* (rev. ed.). New York: Wiley.

Dalgaard, P. (2002). *Introductory statistics with R*. Berlin: Springer.

Deutsch, K. W., & Isard, W. (1961). A note on a generalized concept of effective distance. *Behavioral Science*, *6*, 308–311.

Feenstra, R. C., Rose, A. K., & Markusen, J. R. (2001). Using the gravity model to differentiate among alternative theories of trade. *Canadian Journal of Economics*, *34*, 430–447.

Finley, A. O., Banerjee, S., & Carlin, B. P. (2007, April). spBayes: An R package for univariate and multivariate hierarchical point-referenced spatial models. *Journal of Statistical Software*, *19*(4). Retrieved October 29, 2007, from http://www.jstatsoft.org/v19.

Fotheringham, A. S., Charlton, M., & Brundson, C. (2002). *Geographically weighted regression: The analysis of spatially varying relationships*. New York: Wiley.

Franzese, R. (1999). Partially independent central banks, politically responsive governments, and inflation. *American Journal of Political Science*, *43*, 681–706.

Franzese, R., & Hayes, J. C. (2007). Spatial econometric models for the analysis of TSCS data in political science. *Political Analysis*, *15*, 140–164.

Gartzke, E. (1998). Kant we all just get along? Opportunity, willingness and the origins of the democratic peace. *American Journal of Political Science*, *42*, 1–27.

Geisser, S. (1974). A predictive approach to the random effect model. *Biometrika*, *61*, 101–107.

Geisser, S. (1975). The predictive sample reuse method with applications. *Journal of the American Statistical Association*, *70*, 320–328.

Getis, A., & Boots, B. (1978). *Models of spatial processes*. Cambridge, UK: Cambridge University Press.

Getis, A., & Ord, J. K. (1996). Local spatial statistics: An overview. In P. Longley & M. Batty (Eds.), *Spatial analysis: Modelling in a GIS environment* (pp. 261–277). Cambridge, UK: Geoinformation International.

Geyer, C. J., & Thompson, E. A. (1992). Constrained Monte Carlo, maximum likelihood for dependent data (with discussion). *Journal of the Royal Statistical Society, Series B, Methodological*, *54*, 657–699.

Gleditsch, K. S. (2002a). *All international politics is local: The diffusion of conflict, integration, and democratization*. Ann Arbor: University of Michigan Press.

Gleditsch, K. S. (2002b). Expanded trade and GDP data. *Journal of Conflict Resolution*, *46*, 712–724.

Gleditsch, K. S., & Ward, M. D. (1997). Double take: A re-examination of democracy and autocracy in modern polities. *Journal of Conflict Resolution*, *41*, 361–382.

Gleditsch, K. S., & Ward, M. D. (2000). War and peace in time and space: The role of democratization. *International Studies Quarterly*, *44*, 1–29.

Gleditsch, K. S., & Ward, M. D. (2001). Measuring space: A minimum distance database and applications to international studies. *Journal of Peace Research*, *38*, 749–768.

Gleditsch, K. S., & Ward, M. D. (2007). The diffusion of democracy and the international context of democratization. *International Organization*, *60*, 911–933.

Grenander, U. (1954). On the estimation of regression coefficients in the case of autocorrelated disturbance. *Annals of Mathematical Statistics*, *25*, 252–272.

Griffith, D. A. (1996). Some guidelines for specifying the geographic weights matrix contained in spatial statistical models. In S. Arlinghaus (Ed.), *Practical handbook of spatial statistics* (pp. 65–83). Boca Raton, FL: CRC Press.

Griffith, D. A. (2003). Using estimated missing spatial data with the 2-median model. *Annals of Operations Research*, *122*, 233–247.

Haining, R. (2003). *Spatial data analysis: Theory and practice* (1st ed.). Cambridge, UK: Cambridge University Press.

Holmes, T. J. (2006, February). *Geographic spillover and unionism.* National Bureau of Economic Research (Working Paper Series 12025). Retrieved October 17, 2007, from http://www.nber.org/papers/w12025.

Hubert, L. J., Golledge, R. G., & Constanzo, C. M. (1981). Generalized procedures for evaluating spatial autocorrelation. *Geographical Analysis, 12*, 224–233.

Huffer, F. W., & Wu, H. (1998). Markov chain Monte Carlo for autologistic, regression models with application to the distribution of plant species. *Biometrics, 54*, 509.

Imai, K. (2005). Do get-out-the-vote calls reduce turnout? The importance of statistical methods for field experiments. *American Political Science Review, 99*, 283–300.

Işik, O., & Pinarcioğlu, M. M. (2007). Geographies of a silent transition: A geographically weighted regression approach to regional fertility differences in Turkey. *European Journal of Population, 22*, 399–421.

Jaggers, K., & Gurr, T. R. (1995). Tracking democracy's "Third Wave" with the Polity III data. *Journal of Peace Research, 32*, 469–482.

Jin, X., Banerjee, S., & Carlin, B. P. (2007). Order-free coregionalized areal data models with application to multiple disease mapping. *Journal of the Royal Statistical Society, Series B, 69*, 817–838.

Johnson, S. (2006). *The ghost map.* New York: Riverhead Books.

Jones, D. M., Bremer, S. A., & Singer, J. D. (1996). Militarized interstate disputes, 1816–1992: Rationale, coding rules, and empirical applications. *Conflict Management and Peace Science, 15*, 163–213.

Keele, L., & Kelly, N. J. (2006). Dynamic models for dynamic theories: The ins and outs of lagged dependent variables. *Political Analysis, 14*, 186–205.

Kenny, D. (1981). Interpersonal perception: A multivariate round robin analysis. In M. B. Brewer & B. E. Collins (Eds.), *Scientific inquiry and the social sciences: A volume in honor of Donald T. Campbell* (pp. 288–309). San Francisco: Jossey-Bass.

Kidron, M. (1981). *The state of the world atlas.* New York: Simon & Schuster.

Lacombe, D. (2004). Does econometric methodology matter? An analysis of public policy using spatial econometric techniques. *Geographical Analysis, 36*, 105–118.

Leamer, E. E. (1978). *Specification searches: Ad hoc inference with non-experimental data.* New York: Wiley.

Lee, C.-S. (2005). Income inequality, democracy, and public sector size. *American Sociological Review, 70*, 158–181.

Leontief, W. W. (1986). *Input-output economics.* New York: Oxford University Press.

Lin, T.-M., Wu, C.-E., & Lee, F. Y. (2006). Neighborhood influence on the formation of national identity in Taiwan: Spatial regression with disjoint neighborhoods. *Political Research Quarterly, 59*, 35–46.

Lipset, S. M. (1959). Some social requisites of democracy. *American Political Science Review, 53*, 69–105.

Lofdahl, C. (2002). *Environmental impacts of globalization and trade: A systems study.* Cambridge: MIT Press.

Malloy, T. E., & Kenny, D. A. (1986). The social relations model: An integrative method for personality research. *Journal of Personality, 54*, 199–225.

Matheron, G. (1963). Principles of geostatistics. *Economic Geology, 58*, 1246–1266.

Moran, P. A. P. (1950a). Notes on continuous stochastic phenomena. *Biometrika, 37*, 17–23.

Moran, P. A. P. (1950b). A test for serial independence of residuals. *Biometrika, 37*, 178–181.

Morrow, J. D., Siverson, R. M., & Tabares, T. E. (1998). The political determinants of international trade: The major powers, 1907–90. *American Political Science Review, 92*, 649–661.

Murdoch, J. C., Sandler, T., & Sargent, K. (1997). A tale of two collectives: Sulfur versus nitrogen oxides emission reduction in Europe. *Economica, 64*, 281–301.

Ord, J. K. (1975). Estimation methods for models of spatial interactions. *Journal of the American Statistical Association, 70*, 120–126.

Ord, J. K., & Getis, A. (1995). Local spatial autocorrelation statistics: Distributional issues and an application. *Geographical Analysis, 27*, 286–306.

Pollins, B. M. (1989a). Conflict, cooperation, and commerce: The effect of international political interactions on bilateral trade flows. *American Journal of Political Science, 33*, 737–761.

Pollins, B. M. (1989b). Does trade still follow the flag? A model of international diplomacy and commerce. *American Political Science Review, 83*, 465–480.

R Development Core Team. (2004). *R: A language and environment for statistical computing*. Vienna, Austria: R Foundation for Statistical Computing (ISBN 3-900051-00-3; http://www.R-project.org).

Ripley, B. D. (1981). *Spatial statistics*. New York: Wiley.

Ripley, B. D. (1988). *Statistical inference for spatial processes*. Cambridge, UK: Cambridge University Press.

Rose, A. K. (2004). Does the WTO really increase trade? *American Economic Review, 94*, 98–114.

Rozanski, J., & Yeats, A. (1994). On the (in)accuracy of economic observations: An assessment of trends in the reliability of international trade statistics. *Journal of Development Economics, 44*, 103–130.

Rue, H., & Held, L. (2005). *Gaussian Markov random fields: Theory and applications*. London: Chapman & Hall.

Schabenberger, O., & Gotway, C. A. (2005). *Statistical methods for spatial data analysis*. Boca Raton, FL: Chapman & Hall.

Shin, M. E. (2001). The politicization of place in Italy. *Political Geography, 20*, 331–352.

Shin, M. E., & Agnew, J. (2002). The geography of party replacement in Italy, 1987–1996. *Political Geography, 21*, 221–242.

Shin, M. E., & Agnew, J. (2007a). *Berlusconi's Italy: Where it started, where it ended*. Philadelphia: Temple University Press.

Shin, M. E., & Agnew, J. (2007b). The geographical dynamics of Italian electoral change, 1987-2001. *Electoral Studies, 26*, 287–302.

Signorino, C., & Ritter, J. (1999). Tau-b or not tau-b. *International Studies Quarterly, 43*, 115–144.

Tiefelsdorf, M. (1972). The saddlepoint approximation of Moran's I and local Moran's I: Reference distributions and their numerical evaluation. *Geographical Analysis, 34*, 187–206.

Tufte, E. R. (1990). *Envisioning information*. Cheshire, CT: Graphics Press.

Tufte, E. R. (1992). *The visual display of quantitative information*. Cheshire, CT: Graphics Press.

Tufte, E. R. (1997). *Visual explanations: Images and quantities, evidence and narrative*. Cheshire, CT: Graphics Press.

Varian, H. R. (1972). Benford's law. *American Statistician, 26*, 65.

Wainer, H. (2004). *Graphic discovery: A trout in the milk and other visual adventures*. Princeton, NJ: Princeton University Press.

Wall, M. M. (2004). A close look at the spatial structure implied by the CAR and SAR models. *Journal of Statistical Planning and Inference, 121*, 311–324.

Waller, L. A., Carlin, B. P., Xia, H., & Gelfand, A. E. (1997). Hierarchical spatio-temporal mapping of disease rates. *Journal of the American Statistical Association, 92*, 607–617.

Ward, M. D., & Gleditsch, K. S. (2002). Location, location, location: An MCMC approach to modeling the spatial context of war and peace. *Political Analysis, 10*, 244–260.

Ward, M. D., & Hoff, P. D. (2007). Persistent patterns of international commerce. *Journal of Peace Research, 44*, 157–175.

Ward, M. D., Siverson, R. M., & Cao, X. (2007). Disputes, democracies, and dependencies: A reexamination of the Kantian peace. *American Journal of Political Science, 51*, 583–601.

Wasserman, S., & Faust, K. (1994). *Social network analysis: Methods and applications*. Cambridge, UK: Cambridge University Press.

Watts, D. J. (2003). *Six degrees: The science of a connected age*. New York: W. W. Norton.

West, W. J. (2005). Regional cleavages in Turkish politics: An electoral geography of the 1999 and 2000 national elections. *Political Geography, 24*, 499–523.

You, J.-S., & Khagram, S. (2005). A comparative study of inequality and corruption. *American Sociological Review, 70*, 136–157.

INDEX

ABOUT THE AUTHORS

Michael D. Ward is Professor of Political Science at the University of Washington, Seattle, Washington (http://faculty.washington.edu/mdw) where he teaches courses in political methodology and international relations. He is an affiliate of the Center for Statistics and the Social Sciences. His current work involves the examination of dependencies in dynamic networks of international trade and conflict. A second major project focuses on the social distances among different national groups in Bosnia and the North Caucasus region of Russia.

Kristian Skrede Gleditsch is Professor in the Department of Government at the University of Essex (http://privatewww.essex.ac.uk/~ksg) and a research associate of the Centre for the Study of Civil War at the International Peace Research Institute, Oslo (PRIO). His research interests include conflict and cooperation, democratization, and spatial dimensions of social and political processes.